Solar Energy Markets

Solar Energy Markets
An Analysis of the Global Solar Industry

Philip G. Jordan

The Economic Advancement Research Institute
Wrentham, MA, USA

and

BW Research Partnership, Inc.
Wrentham, MA, USA

AMSTERDAM • BOSTON • HEIDELBERG • LONDON • NEW YORK • OXFORD
PARIS • SAN DIEGO • SAN FRANCISCO • SINGAPORE • SYDNEY • TOKYO

Elsevier
32 Jamestown Road, London NW1 7BY, UK
225 Wyman Street, Waltham, MA 02451, USA

British Library Cataloguing-in-Publication Data
A catalogue record for this book is available from the British Library

Library of Congress Cataloging-in-Publication Data
A catalog record for this book is available from the Library of Congress

ISBN: 978-0-12-397174-6

For information on all Elsevier publications
visit our website at store.elsevier.com

This book has been manufactured using Print On Demand technology. Each copy is produced to order and is limited to black ink. The online version of this book will show color figures where appropriate.

Contents

Acknowledgments ix

1 **Introduction: An Overview of the Solar Industry** 1

2 **The Mechanics of Solar Power** 7
- Installed Capacity 7
  - Photovoltaic 7
- The Manufacturing Process 9
- Thin Film 10
- Concentrated Solar Energy 11
  - Concentrated Photovoltaics 11
  - Solar Thermal Energy 12
- Parabolic Trough 14
- Power Tower 15
- Fresnel Reflectors 16

3 **The New Culture of Environmentalism** 19
- Global Trends 20
- United States 21
- Europe 24
- China 25

4 **Finance and Venture Capital** 29
- Research, Development, and Deployment 29
- Private Investment Trends 31
- Global Investments 31
  - Comparative Investments 33
  - United States 36
  - Germany 37
  - Spain 38
  - Italy 39
  - India 40
- Public Sector Spending 42

5 **Global Solar Policy** 43
- Energy Policy 43
- EU Policy Framework 43

EU Supply-Side Policies 44
Germany 45
German Supply-Side Policies 46
German Economic Development Strategies 48
German Demand-Side Policies 51
Spanish Supply-Side Policies 51
Spanish Economic Development Strategies 52
Spanish Demand-Side Policies 53
Summary 53
Italy 54
Italian Supply-Side Policies 54
Italian Economic Development Strategies 58
China 59
Chinese Supply-Side Policies 59
Chinese Economic Development Strategies 61
Chinese Demand-Side Policies 62

6 Federal and State Energy Policies 65
Federal Policies 65
Tax Credits 65
Modified Accelerated Cost Recovery System 66
Grants 68
Loans and Loan Guarantees 69
Department of Energy Loan Guarantee Program 69
Clean Renewable Energy Bonds 69
Qualified Energy Conservation Bonds 69
State Policies 70
Renewable Portfolio Standards 71
Tax Credits 72
Direct Cash Financing 85
PACE 95
Property Tax Incentives 98
Economic Development Incentives 101
Permitting 103
Loan Programs 104

7 The Solar Labor Market—Efficiencies and Productivity 111
United States 112
Installation 116
Manufacturing 119
Sales and Distribution 121
Project Developers 122
Other 122
Germany 123
China 125

8 Global Markets **127**
Global Energy Demand **127**
Residential **128**
Commercial **129**
Industrial **129**
Global Renewable Energy Outlook **130**

9 The Economics of Solar Power **135**

10 Afterward **143**

Glossary **145**

Acknowledgments

I would like to offer my very special thanks to my colleague and friend Dr. Edward A. Cunningham. His advice has been a great help in the development of this book.

I am particularly grateful to Barry Friedman of the National Renewable Energy Laboratory, Andrea Luecke of The Solar Foundation, and Tom Kimbis and Justin Baca from SEIA for their deep understanding of the importance of solar labor market issues and the assistance that they have given me in developing my career.

I also wish to acknowledge Ryan Young, our research analyst at BW Research Partnership, who provided exemplary data collection and analysis for this text.

I would also like to express my deep gratitude to Professor Zygmunt Plater and Professor Benjamin Sachs. Each played a pivotal role in my professional development and instilled within me a deep drive to discover, to learn, to teach, and to act.

This book would not have been possible without the constant, patient support that I received from my wife, Anne. Thank you for the late night read-throughs and gentle nudges along the way.

1 Introduction: An Overview of the Solar Industry

The 2007 US banking crisis was the first of a series of shocks to the global economy with trillions of dollars of wealth evaporating from the globe and years of scandal and upheaval to follow. Economic markets have been slow to recover, and labor markets even slower. Few industries have grown, let alone thrived.

Despite these unprecedented downward global pressures, *the solar industry has experienced a global revolution* with profound implications for business, government, and the environment. This text is intended to provide a unique, global perspective of the US solar industry, exploring the differences between the solar industry today and previous growth spurts such as the 1970s brief solar boom.

This book relies on information from the nation's first comprehensive solar industry survey, pioneering survey work from adjacent industries, and insights from key thought leaders in the energy sector in the United States, and from international leaders in solar development.

At its most basic level, solar power is a broadly defined term for harnessing the power of the sun to generate heat or electricity, which humans have been doing for at least 5000 years! There are many different technologies at use in the commercially available products that capture solar energy, from passive design features of buildings to advanced thin film photovoltaic (PV) panels. Each of these products share similarities, such as shared incentives and market drivers, as well as many differences from their technological innovations to their economic viability.

Generally speaking, the solar industry is categorized by the various technology-driven product lines that make up the array of choices for the global consumer. The three largest categories by technology are PV (electric power generation), solar thermal (electric power generation), and solar water heating. Solar space heating and cooling are also growing areas with particularly strong potential in the northeast United States, though such applications are clearly well behind the other uses in terms of market penetration.

Solar thermal products use solar energy to heat water or other liquids. These can be used for heating water for domestic/commercial use or to produce electricity through the use of a steam turbine system. Solar water heaters employ a simple design utilizing aluminum fins and insulated storage tanks to supply hot water for pools or domestic use. To generate electricity, however, much more heat is needed, and the most common mechanism for obtaining this heat is through concentrating solar power (CSP). CSP uses mirrors to focus solar rays to provide intense heat that generate significant steam, which can then be passed through a variety of steam turbine systems.

The majority of this book will focus on the largest segment photovoltaics, but will also include pertinent details in each chapter regarding solar thermal technologies.

Solar Energy Markets. DOI: http://dx.doi.org/10.1016/B978-0-12-397174-6.00001-5

Due to recent price declines in photovoltaics and the much lower maintenance required (PV panels have no moving parts!), many projects throughout the southwest United States that were planned to use CSP have been changed to PV projects.

Across the globe, recent significant price drops in traditional PV panels have significantly changed the solar industry, shifting interest away from producing more efficient products towards producing traditional photovoltaics even more efficiently. With price declines of approximately 70% over a 2-year period, the economics of PV power systems have improved dramatically and far outcompete rival solar technologies.

PV products represent the lion's share of the solar industry. In a recent survey of solar employers in the United States, over 90% of all solar installation companies work with PV products.[1] Photovoltaics operate by using arrays of semiconductors, typically made of monocrystalline or polycrystalline silicon, to produce direct current (DC) energy from solar radiation.[2] The global rise in PV panel installation has led to significant growth of solar electric power in Europe and the United States, and dramatic increases in panel manufacturing throughout the globe.[3]

The United States has installed approximately 4 GW of solar power through 2011, tying it with Spain for fourth place in total generation, behind Germany, Italy, and Japan.[4] Though the total amount of energy generated by photovoltaics has increased dramatically over the last 10 years, the overall demand growth over the first part of that period has meant that PV merely kept up with other technologies, as the percentage of electricity produced by PV systems had not changed significantly over time.[5]

This inability to capture increasing share of the electrical profile shifted in 2009, when the solar capacity of the United States experienced incredible growth, with no signs of a slowdown. During this solar boom, for the first time in generations, the United States experienced energy demand *declines* due to the great recession and accompanying slow recovery. And the pace continued to quicken; utility-driven PV installations increased 109% alone in 2011 representing an additional 758 MW of solar power.[6]

Installations only tell part of the US solar story. Despite widespread misconceptions, perpetuated by media stories, the United States is a net exporter of solar products, meaning that US manufacturers produce more solar components than are installed domestically. Historically, only 30% of the photovoltaics installed in this county are domestically sourced, but the United States exports large quantities of solar products to other nations. In 2010, for example, the United States imported $3.7 billion of solar products, while exporting $5.6 billion, resulting in a net export of nearly $2 billion in the industry.[7]

Every mainstream discussion on photovoltaics eventually leads to China, but it is becoming increasingly apparent that it is for the wrong reasons. Though it is true that Chinese contribution to the global industry has been primarily related to production

[1] The Solar Jobs Census 2011. The Solar Foundation, October 2011.

[2] See http://science.nasa.gov/science-news/science-at-nasa/2002/solarcells/.

[3] http://www.eia.gov/cneaf/solar.renewables/page/solarphotv/solarpv.html.

[4] BP Statistical World Energy Review 2011 (retrieved 8.08.11). EurObserver 202: Photovoltaic Barometer.

[5] http://www.eia.gov/totalenergy/data/annual/showtext.cfm?t=ptb1008.

[6] SEIA and GTM Research, March 12, 2012.

[7] *id.*

(as the largest producer of solar products), it is becoming increasingly apparent that demand-pull from China will be the single most important factor shaping the future of the solar industry in the United States, potentially remaking the economics and labor force of the US solar industry.

Currently, increased production of low-cost Chinese panels has resulted in significant price declines for global PV installations. In 2011, prices dropped by an incredible 30%, as part of a 70% decline over the last 30 months.[8] This price drop has clearly negatively impacted manufacturers outside of China, prompting a trade complaint filed (and won, at least temporarily) by US manufacturers.

Equally apparent, however, is that the price declines have spurred the US installation market. Together with strong federal and local incentive programs, the low price of equipment has led to significant global increases in solar installations, leading many experts to believe that PV-produced energy will soon reach price parity—some believing as early as 5–10 years from now.[9]

Price parity, the elusive holy grail of the industry, will have as much (or more) to do with China than perhaps any other region. On the one hand, and as previously mentioned, module price declines from Chinese manufacturers have brought PV power dramatically closer to parity. Declining cost trajectory would obviously hasten this trend, however, China's direction in terms of installations will likely be the key to prices in the future.

Like any commodity, supply *and* demand dictate pricing, and the future of Chinese demand has as much to do with forecasting prices as does the supply output. As of 2012, China has installed approximately 7 GW of solar power, but it has set a goal more than doubling that by adding 10 GW of solar power in 2013 alone.[10] As reported in Reuters in January of 2013, this sets China on a strong path to achieve their previously stated goal of 21 GW of solar power by 2015.

As noted in that report, however, this alone is not sufficient to significantly drive prices up or spur greater innovation for future market response. According to Morningstar Analyst Stephen Simko (as reported by Reuters), "If you look at how much supply there is in the world relative to demand, even if China grows by 10 GW this year, it really is not enough to fix the problems that exist in the solar sector…"[11]

A persistent and sustained increase in capacity additions in China together with continued growth in Europe and the United States would likely bring supply to a level that would increase prices in the short term. This spike is inevitably followed by greater manufacturing innovation and efficiencies, which lead to permanent price declines. As a result, one likely pathway for sustained price parity with fossil generation includes temporary price spikes due to rapid expansion of installed capacity.

Germany has the world's most mature solar market with nearly 28 GW of installed solar capacity in 2012, adding nearly 7.5 GW in 2011 alone. Coupled with its 18 TWh of thermal energy, solar power contributes 3% of Germany's overall

[8] See http://www.reuters.com/article/2012/04/13/solar-prices-idUSL2E8FAD0 × 20120413.

[9] See Lorenz, A. 1366 Technologies, speaking at the 2012 MIT Energy Conference; Lovins, A. Reinventing Fire: Bold Business Solutions for the New Energy Era, 2012.

[10] http://www.pv-magazine.com/news/details/beitrag/china-sets-10-gw-installed-capacity-target-for-2013_100009803/#axzz2VMqehJJD.

[11] *id.*

output, an increase from 0.01% in 2000. Germany presents a compelling example of how to build a solar industry in a large country with a significant economic and manufacturing base.

Throughout this text, each chapter will include a topical overview for the global industry with specific emphasis on the US markets. In addition, trends in Germany and China will provide additional detail and serve as reference points for comparison. In this way, three of the major solar markets will be covered in detail.

Any analysis of the economic or workforce implications of solar energy must delineate not only by technology (and to some extent, geography) but also by scope. Distributed generation, or production of electricity at the site of consumption, includes projects that are much smaller in size and scope than utility-scale projects, where power is produced in mass quantity and delivered to customers through the grid. Much more labor intensive, distributed generation maintains its cost competitiveness because there is virtually no loss in transmission and in most cases, no payments or fees to the electric utilities. In fact, in many regions of the country, grid-tied distributed generation system owners receive credits or payments from their utilities for their surplus power production that can be transmitted to other customers.

Solar installations are therefore generally segmented into three categories: (1) residential, such as homeowner, rooftop solar; (2) nonresidential, such as commercial building or campus-wide systems; and (3) utility-scale, large systems designed specifically for feeding the grid rather than specific uses.

In addition to price declines in manufacturing and installation, low interest rates, beneficial tax policies, state renewable energy credits, and new financing models, solar power is rapidly approaching price parity with traditional, fossil-fuel electric prices in the retail market. Despite the many economic and social benefits of solar power, however, consumers have responded more slowly than expected and most of the new solar megawatts installed in the United States come from commercial and utility-scale projects.

Recent estimates indicate that the United States installed 1.85 GW of PV power in 2011. These new additions were led by commercial installations of about 800 MW, followed closely by utility-scale projects (758 MW) and 297 MW of residential PV generation.[12]

Some of this is likely due to concerns about the boom–bust history of the solar industry, particularly during the 1970 energy crisis. The crisis and accompanying oil embargoes spurred America to action to develop and refine technologies to generate power.

Current solar panel designs were developed in the late 19th and early 20th centuries (both thermal and PV), but they were very expensive to produce. Around the same time that the energy crisis was in swing, new technologies brought the price down from nearly $100 per watt to $20. The price declines and political will for a sustainable source of domestic energy led to late 1970s production exceeding 500 kW of solar power and widespread use from oil rigs to spacecraft, and all manner of off-grid applications in between. By 1983, solar projects were generating more than 21 MW of energy, and the solar industry was a $250 million per year business.

[12] SEIA and GTM Research, March 12, 2012.

The mid-1980s brought an end to the energy crisis and oil dipped below $15 per barrel. American consumers resumed their pre-crisis consumption patterns, and political leaders heeded the changing winds and ended or dramatically reduced subsidies. The industry that looked to have such promise was, for all intents and purposes, dead.

This boom–bust cycle has led many to be cynical, suggesting that the solar industry will yet again fail to live up to expectations. There are significant differences between the solar industry's current position and its past, however, which make it unlikely that—despite recent headlines regarding Solyndra and other high-profile bankruptcies—history will repeat itself in this instance.

This book examines the key drivers of why the present success of the solar industry differs from past experience by examining six key drivers of the solar industry in a global context: (1) a new culture of environmentalism, (2) policy and markets, (3) financing and venture capital, (4) economics and cost competitiveness, (5) innovation, and (6) labor.

2 The Mechanics of Solar Power

A basic primer on the mechanics of solar power generation is important in order to develop the foundational knowledge required to properly consider the economic and workforce trends of the solar industry. This chapter provides an overview of how each of the various technologies (PV, thermal, etc.) work. It is neither as exhaustive as a scientific textbook on the subject nor is it as simple as a glossary entry; rather it provides the introductory knowledge base that is a prerequisite for analyzing and considering the industry as a whole.

Installed Capacity

Global Installed Totals (MWp)

	2008	2009	2010	2011	2012
CSP[a,b]	484	663	969	1598	2553
PV[c]	16,200	23,600	40,700	71,100	102,200

[a]CSP Facts & Figures. Csp-world.com (retrieved 22.04.13).
[b]Concentrating Solar Power. irena.org, p. 11.
[c]European Photovoltaic Industry Association, 2013. Global Market Outlook for Photovoltaics 2013–2017.
Source: CSP: https://en.wikipedia.org/wiki/Concentrated_solar_power; PV: http://en.wikipedia.org/wiki/Photovoltaics.

Photovoltaic

At its most basic level, solar PV technology converts sunlight to energy. This conversion happens directly through solar cells made up of various components that produce the photoelectric effect. This phenomenon occurs when electrons are emitted from materials as they absorb light energy. Though early PV cells were made of silver selenide or copper, silicon has been the predominant substance used for the past 60 years.[1]

Total PV Peak Power Capacity (MWp)[a]

Country or Region	2010	2011	2012
World	39,778	69,684	102,024
European Union	29,328	51,360	–
Germany	17,320	24,875	32,509
Italy	3502	12,764	16,987
China	893	3093	8043
United States	2519	4383	7665

(*Continued*)

[1]http://www.eere.energy.gov/basics/renewable_energy/types_silicon.html.

Solar Energy Markets. DOI: http://dx.doi.org/10.1016/B978-0-12-397174-6.00002-7

Total PV Peak Power Capacity (MWp)[a] (*Continued*)

Country or Region	**2010**	**2011**	**2012**
Japan	3617	4914	6704
Spain	3892	4214	–
France	1025	2831	3843
Belgium	803	2018	–
Czech Republic	1953	1960	–
Australia	504	1298	2291
United Kingdom	72	1014	1831
India	189	461	1686
South Korea	662	754	–
Greece	206	631	–
Canada	200	563	–
Slovakia	145	488	–
Switzerland	111	216	–
Israel	66	196	–
Ukraine	3	190	–
Austria	103	176	–
Portugal	131	144	–
Bulgaria	18	133	1066
Netherlands	97	118	–
Taiwan	32	102	–
Slovenia	36	90	–
South Africa	40	41	–
Mexico	30	40	–
Brazil	27	32	–
Luxembourg	27	31	–
Sweden	10	19	–
Denmark	7.1	17	–
Malaysia	15	15	–
Finland	9.6	11	–
Cyprus	6.2	10	–
Norway	9.2	9.2	–
Turkey	6	6	–

[a]BP Statistical World Energy Review 2011 (XLS). EurObserv'ER 202: Photovoltaic Barometer (retrieved 8.08.11).
Source: http://en.wikipedia.org/wiki/Solar_power_by_country.

Solar cells produce DC power, which fluctuates based on the sun's intensity. This is why cloud cover, seasonal angle of the sun, and therefore latitude, and new solar tracking systems have such dramatic impacts on generation. Typically, cells are cut from large segments of bulk material called wafers and processed as semiconductors.

These cells are connected together to form modules. A module is what most people would identify as a solar panel. Modules are connected to form arrays, which would represent the aggregate of panels on a rooftop installation. Finally, in order to be connected to a grid, inverters convert the DC to alternating current (AC).

The Manufacturing Process

Much of solar PV systems' current price structure is based on the supply of materials and the efficiency of the manufacturing process. Other important considerations, covered later in this book, are consumer demand, government policies and incentives, the availability and cost of capital, raw materials and component pricing, and labor availability and wages.

The solar manufacturing process begins with raw materials and in the case of most traditional panels that material is either monocrystalline or polycrystalline silicon. Silicon is mined, typically from sand or sandstone, and is the second most abundant material on earth. Any grade school student could easily identify it as quartz!

The mined material, silicon dioxide, is readily available, yet must go through an extensive purification process to be used in solar applications. Purification is conducted in multiple steps. First, the material is heated in a furnace to release the oxygen and separate the carbon dioxide from the molten silicon. This 99% pure silicon is then "purified even further using the floating zone technique. A rod of impure silicon is passed through a heated zone several times in the same direction. This procedure "drags" the impurities toward one end with each pass. At a specific point, the silicon is deemed pure and the impure end is removed."[2]

Using the *Czochralski method*, a crystal of silicon is used as a seed, dropped into polycrystalline silicon, and a large, pure, cylindrical ingot of silicon results. Ingots of raw materials are, in the case of polycrystalline cells, cut into wafers that are thin slices that are polished and treated with doping agents (as are most semiconductors). Doping agents are simply impurity atoms that create either p-type which has extra holes or n-type which contains excess free electrons. The junctions of these regions are called p–n junctions and their purpose is to increase the conductivity of the cell.

Cells then receive metal contacts and are connected to other cells to create modules. These contacts are typically joined with tin and copper connectors.

The cells then go through several processes to reduce reflectivity and increase absorption of light. Nearly all cells receive an antireflective chemical coating, which is typically silicon nitride. Also, some cells are textured for to reduce reflective properties.

Finally, modules are encapsulated in a silicon- or vinyl-based compound and cased in aluminum or other lightweight metal frames. These frames typically receive construction grade Mylar backing and are capped with glass.

Given these complex processes, the finished product assembly as illustrated above only represents a small fraction of the solar manufacturing industry. Component manufacturers in photovoltaics may be involved with the production of

- Ingots
- Wafers
- Cells
- Modules
- Inverters
- Racking

[2]http://www.madehow.com/Volume-1/Solar-Cell.html.

- Glass
- Laminates
- Materials/mining.

Thin Film

While traditional PV mono- or polycrystalline cells make up the majority of the PV market, thin film solar is gaining ground, particularly in the United States. Thin film technology has a multitude of benefits and chief among them is lower cost. However, the transition to thin film has not been smooth, despite the technology having been in use in calculators and watches for decades, and significant improvements in efficiency must be gained before thin film encroaches on traditional PV's market share.

At about 1/350th the thickness, thin film solar panels are obviously manufactured quite differently from traditional PV cells. Rather than the material-intensive identified above, thin film is produced by applying single layers of semiconductors on a substrate—most typically glass, or sometimes metal or plastic. This thin sheet is lighter, cheaper, yet less efficient than the thick wafer cells.

Thin film technology has been around for decades. Solar-powered calculators and other small devices have used amorphous silicon (a-Si) thin film to charge their tiny batteries for decades. However, in large-scale applications, researchers have been unable to increase the efficiencies enough to make a-Si systems on a larger scale more competitive.

However, several alternative technologies are promising—if not yet ready to replace traditional PV systems. These nonsilicon-based technologies have the added benefit of relying on materials that are more readily available and less volatile.

The most promising new thin film technologies use either copper indium gallium deselenide (CIGS) or cadmium telluride (CdTe). There are two basic configurations of the CIGS solar cell: the CIGS-on-glass cell requires a layer of molybdenum to create an effective electrode. This extra layer isn't necessary in the CIGS-on-foil cell because the metal foil acts as the electrode. A layer of zinc oxide (ZnO) plays the role of the other electrode in the CIGS cell. Sandwiched in between are two more layers—the semiconductor material and cadmium sulfide (CdS). These two layers act as the n-type and p-type materials, which are necessary to create a current of electrons.[3]

The newest process for manufacturing thin film solar is best described in the illustration of San Jose, CA based Nanosolar, Inc. manufacturing process.

Nanosolar makes its solar cells using a process that resembles offset printing. Here's how it works:

1. Reams of aluminum foil roll through large presses, similar to those used in newspaper printing. The rolls of foil can be meters wide and miles long. This makes the product much more adaptable for different applications.
2. A printer, operating in an open-air environment, deposits a thin layer of semiconducting ink onto the aluminum substrate. This is a huge improvement over CIGS-on-glass or CdTe

[3] http://science.howstuffworks.com/environmental/green-science/thin-film-solar-cell2.htm.

cell manufacturing, which requires that the semiconductor be deposited in a vacuum chamber. Open-air printing is much faster and much less expensive.
3. Another press deposits the CdS and ZnO layers. The ZnO layer is nonreflective to ensure that sunlight is able to reach the semiconductor layer.
4. Finally, the foil is cut into sheets of solar cells. Sorted-cell assembly, similar to that used in conventional silicon solar technology, is possible in Nanosolar's manufacturing process. That means the electrical characteristics of cells can be matched to achieve the highest panel efficiency distribution and yield. CIGS-on-glass solar panels don't offer sorted-cell assembly. Because their panels consist of cells that are not well matched electrically, their yield and efficiency suffer significantly.[4]

This process is critical because it lowers the price, which is necessary given the market perception of lower overall efficiency of thin film. Traditional solar is reaching about 25% maximum efficiency while CdTe and CIGS are now reaching 15–20%. Further investment in these technologies, as illustrated in Chapter 5, will increase these efficiencies further, drive down manufacturing costs, and increase the feasibility for many applications.

Concentrated Solar Energy

Unlike traditional and thin film PV applications that absorb natural light from the sun, CSP systems work by using mirrors or other reflectors to intensify the sun's rays prior to collection. CSP is growing at a rapid pace due to its feasibility for larger systems—typically utility-scale systems. By the end of 2017, CSP is expected to generate about 10.9 GW of power globally and 4.2 GW in the United States (#2 behind Spain) alone.

The current nomenclature in use to describe concentrating systems can be confusing and quickly becomes riddled with jargon when referring to concentrated photovoltaics (CPV), CSP, and solar thermal energy. A basic explanation is:

- CPV use mirrors to concentrate the PV effect. CPV is generally not included in the discussion of CSP and is clearly not a solar thermal energy.
- Solar thermal refers to generating heat energy from the sun. This heat energy can be used for space heating, water heating (for pools, domestic hot water, or forced hot water heating systems for instance), or electrical production (typically CSP).
- CSP refers to a variety of technologies that transfer heat energy from the sun to generate electricity.

A summary of each is described in the following sections.

Concentrated Photovoltaics

As its name suggests, CPV uses mirrors or lenses to concentrate light to a smaller but more intense beam. This concentration allows for fewer, smaller panels. Because fewer panels are required, more costly, higher efficiency panels are feasible for the application.

[4] http://science.howstuffworks.com/environmental/green-science/thin-film-solar-cell3.htm.

Despite this cost savings, CPV has added costs that are not typical of a traditional PV system. These costs include the concentrating medium (mirrors, glass, lenses, etc.), tracking equipment, and cooling equipment due to the high heat generated from the concentrated rays. Despite capturing only a small fraction of the overall solar market, forecasts for the technology are strong.

Capital costs for CSP plants continue to fall, making the technology more competitive with fossil-fuel generation. In fact, capital costs for CSP plants without thermal energy storage can be as low as $4600/kW, while plants with between 6 and 15h of thermal storage can carry capital costs as high as $10,500/kW.[5] This point is critical because it is becoming evident that CSP can only compete with PV on its storage potential.[6]

A recent article by Dino Green refines the point, stating: "energy investors consider **competitive cost of energy** the **most important issue**. That is why in 2011 in the US we have seen a sudden shift from planned CSP power plants being converted to Photovoltaic (PV) – this trend continues in 2012. As long as energy price of PV plants is less than the Energy price of equivalent CSP, and continue to decline, PV will remain a preferable solution over CSP for energy investors. CSP systems will need to demonstrate high performance in all three attributes, *competitive thermal-energy-storage costs*, *energy dispatch-ability* and *reliability as an ancillary solution*, in order to remain attractive and competitive against Photovoltaic panels."[7]

This is particularly important given PVs declining installed cost per watt and much cheaper operation and maintenance costs, as these costs for CSP plants range from $0.02 to $0.035/kWh.[8]

Solar Thermal Energy

Solar thermal energy refers to heating water with the sun, a practice that has been used by humanity for thousands of years. As the technology has progressed, the United States Energy Information Administration has classified the collectors as low, medium, and high temperature, obviously based on heat output. Generally speaking, low-temperature collectors are used for heating swimming pools or for solar heating and cooling, which utilized heat pump technology as part of a comprehensive Heating, Ventilation, and Air-Conditioning (HVAC) system. Because of their limited use, this text does not address the technology in detail.

Medium-temperature collectors, on the other hand, make up the vibrant and growing solar water heating segment of the industry that supplies hot water for residential and commercial applications. These collectors are able to generate significantly more heat, which is important for applications that require heating to 125–140°F (as opposed to 70–80°C for a pool).

Solar water heating has been shown to be an effective means to supplement traditional water heaters, particularly in southern climates—though their application, with

[5] http://www.irena.org/DocumentDownloads/Publications/RE_Technologies_Cost_Analysis-CSP.pdf.

[6] http://www.renewableenergyworld.com/rea/blog/post/2013/03/how-solar-pv-is-winning-over-csp.

[7] http://www.renewablegreenenergypower.com/solar-energy-facts-concentrated-solar-power-csp-vs-photovoltaic-pv-panels/.

[8] http://www.irena.org/DocumentDownloads/Publications/RE_Technologies_Cost_Analysis-CSP.pdf.

a few tweaks, is effective in even the northernmost reaches of the United States. In the United States, solar water heating is recovering from a bad reputation, mostly driven from poor installation quality in the 1970s in California. However, solar water heating systems are used widely throughout Australia, Europe, Asia (and in particular, Japan), and the Middle East. China is by far the fastest growing location for solar water heating installations representing between 60% and 80% of annual installations globally.

Solar water heating is much more effected by climate than PV, as ambient temperature, overheating and freezing protection, and other consideration can dramatically impact required and actual output for the systems.

The type, complexity, and size of a solar water heating system are mostly determined by

- Changes in ambient temperature and solar radiation between summer and winter.
- The changes in ambient temperature during the day–night cycle.
- The possibility of the potable water or collector fluid overheating.
- The possibility of the potable water or collector fluid freezing.[9]

The major reason for this is to protect the systems (and users) from overheating and freezing. While there are many technologies that can allow for draining or passive heat loss (typically at night) for direct systems, one important technological breakthrough is the development of indirect systems.[10]

The most basic technology for solar water heating is the Integrated Collector Storage (ICS) or batch collection system. The collectors are markedly low tech and inexpensive, and basically serve as water tanks inside an insulated oven that is heated by the sun. Though generally low efficiency, they can be used effectively in warm, sunny regions with less heat loss and without freezing temperatures in the winter.

Flat plate solar thermal collectors use pipes (called headers and risers, depending on size and orientation) to increase heating efficiency. These systems are generally used as direct systems (heating potable water) and often used tempered glass to withstand storm debris and hail. Some of these systems include evacuated tube collectors, which use a vacuum (and some different materials) to reduce heat loss. While this reduction can be significant, the evacuated tube collectors (ETC) is less efficient in full sun applications and has had issues with reliability.

Global Cumulative CSP[a,b] and Yearly Installations (MWp)

	2008	2009	2010	2011	2012
Installed	55	179	307	629	803
Cumulative	484	663	969	1598	2553

[a]CSP Facts & Figures. Csp-world.com (retrieved 22.4.13).
[b]Concentrating Solar Power. irena.org, p. 11.
Source: https://en.wikipedia.org/wiki/Concentrated_solar_power.

[9]http://en.wikipedia.org/wiki/Solar_water_heating, citing eere.energy.gov.

[10]Direct systems heat potable water in the collectors and store in a tank. Indirect systems heat a secondary, nonpotable fluid (often propylene glycol) that heats a tank of water.

The highest temperature systems are used for electrical generation, which is known as CSP. CSP systems operate like many other electrical power plants generating power from steam-driven turbines. However, unlike most power plants that generate the steam from fossil fuels or nuclear reactors, CSP systems use focused solar energy to heat the water. As an added benefit, the systems' direct product is heat, which is significantly cheaper and easier to store than electricity. This is important because it allows CSP systems to produce power after the sun goes down.

CSP continues to grow at a record pace despite fierce competition from PV systems. The most recent data at the time of printing suggests that 938 MW of CSP will be commissioned by the end of 2013 in the United States.[11] This is a staggering figure based on current installed CSP, and reflects the global growth of the market, where more than 20 GW of CSP capacity is currently under development worldwide.[12]

Investors may not be as willing to embrace CSP as they were 3–5 years ago, but the long-term trends still seem quite favorable. While PV developers have been able to focus on driving down the cost per watt of their modules, CSP developers have had many different components to work on. As each of these challenges finds a solution, the technology leaps. So much so that, according to IEA's CSP technology roadmap estimates, total installed CSP capacity could reach 337 GW by 2030, tripling to 1089 GW by 2050.[13]

There are four primary technologies of CSP in current use: (1) parabolic trough; (2) power tower; (3) Fresnel reflector; and (4) stirling dish. Given the recent bankruptcy of stirling energy systems (SES), lack of competitiveness with PV systems, and, despite the highest efficiency rating of the four technologies, the future of stirling systems is suspect and is therefore not covered extensively in this text.

Parabolic Trough

Parabolic trough systems uses curved, parabola-shaped reflectors that use mirror coating to concentrate sunlight on a tube filled with liquid. This tube, frequently called a Dewar tube, is usually filled with oil and carries the heated fluid to an engine similar to a traditional power plant.

To reach its maximum thermal efficiency of 60–80%, parabolic reflectors are mounted on tracking systems to follow the sun. The intensity of the concentrated solar rays heats the liquid medium to approximately 400°C.

The future of the technology, in addition to overall component price declines, will depend on improvements in tracking technology. Currently, parabolic trough tracking systems are aligned to the vernal and autumnal equinoxes, meaning that for much of the year (and most dramatically at the summer and winter solstices) the

[11] http://www.seia.org/research-resources/solar-industry-data.
[12] http://www.irena.org/DocumentDownloads/Publications/RE_Technologies_Cost_Analysis-CSP.pdf.
[13] http://www.irena.org/DocumentDownloads/Publications/RE_Technologies_Cost_Analysis-CSP.pdf.

concentration is not at peak. This results in systems that only reach about 1/3 of their maximum theoretical efficiency.

Parabolic Trough Global Applications

Power Plants	**Installed Capacity (MW)**	**Country**	**Developer/Owner**	**Year Completed**
Solana Generating Station	280	United States	Abengoa Solar	2013
Solnova Solar Power Station	150	Spain	Abengoa Solar	2010
Andasol Solar Power Station	150	Spain	ACS Group	2008
Extresol Solar Power Station	150	Spain	Estela Solar	2010
Palma del Rio Solar Power Station	100	Spain	ACCIONA Energy	2010
Manchasol Power Station	100	Spain	INITEC Energía	2011
Valle Solar Power Station	100	Spain	Torresol Energy	2011
Helioenergy Solar Power Station	100	Spain	Abengoa Solar	2011
Aste Solar Power Station	100	Spain	Elecnor/Aries/ ABM AMRO	2012
Shams	100	UAE	Masdar/Abengoa/ Total S.A.	2013

Source: http://en.wikipedia.org/wiki/List_of_solar_thermal_power_stations; http://en.wikipedia.org/wiki/Solar_power_tower.

Power Tower

While the overall principle is the same, the power tower design is completely different from the parabolic trough. Power tower systems, sometimes referred to as heliostat plants, vaguely resemble oil drills, yet produce clean power from the sun.

The technology is fairly basic and not as complex as the parabolic trough. A series of reflectors mounted on tracking systems, called heliostats, reflect sunlight to a central receiver on top of a large tower.[14] This receiver, which typically contains liquid sodium, seawater, or other fluid, is heated to 500–1000°C. Power towers generally offer more efficiency than trough technology, are cheaper due to the use of flat, rather than curved, glass for the reflectors, and can store heat longer. As a result, the National Renewable Energy Laboratory (NREL) estimates that power tower systems should be able to produce electricity at 5.47 cents/kWh by 2020.[15]

Continued price declines are important for the technology, as it lags far behind parabolic trough systems. In fact, of the 1.9 GW of CSP installed through the first quarter of 2012, 1.8 GW were parabolic trough designs.

[14] http://www.solarpaces.org/CSP_Technology/csp_technology.htm.

[15] Assessment of parabolic trough and power tower solar technology cost and performance forecasts (http://www.nrel.gov/solar/parabolic_trough.html).

Power Tower (Heliostat) Global Applications

Power Plants	Installed Capacity (MW)	Country	Developer/Owner	Year Completed
Ivanpah Solar Power Facility	392 (U/C)	United States	BrightSource Energy	2013
Crescent Dunes Solar Energy Project	110 (U/C)	United States	SolarReserve	2013
PS20 Solar Power Tower	20	Spain	Abengoa	2009
Gemasolar	17	Spain	Sener	2011
PS10 Solar Power Tower	11	Spain	Abengoa	2006
Sierra SunTower	5	United States	eSolar	2009
Jülich Solar Tower	1.5	Germany	–	2008
Acme Solar Thermal Tower	2.5	India	–	2012
Beijing Badaling Solar Tower	1.5	China	–	2012
Yanqing Solar Power Station	1	China	–	2010

Source: http://en.wikipedia.org/wiki/List_of_solar_thermal_power_stations.

Fresnel Reflectors

Fresnel reflectors again use mirrors to focus light onto receivers to heat a fluid, which generates electricity via steam engine or heat pump. These mirrors can be curved or flat, and some use parabolic mirrors to enhance the effect. These systems differ from trough and dish systems because they use multiple mirrors focused on a single receiver (therefore allowing many different configurations to maximize their efficiency at different periods of the day and year). They differ from power tower because they operate on a single rather than dual axis.

Fresnel reactors have had a start and stop history, and there are few plants currently generating with the technology. Initially, the significantly lower cost of Fresnel technology led experts to believe it would outcompete trough designs, however, advances in nanotechnology and advanced manufacturing has narrowed the gap.

The most common designs are linear Fresnel reflector (LFR) and compact linear Fresnel reflectors (CLFR). These systems lower cost and increase output by using absorbers that share mirrors. The principal design flaw, however, is that shading is produced from adjacent reflectors, lowering the overall output of the system. This concern is reduced in CLFR technology, which uses more advanced tracking and situates components closer to the ground. CLFR systems typically use flat mirrors, which are also less expensive to produce, and uses multiple absorbers, maximizing the reflected light.

Fresnel reflectors utilize the Fresnel lens effect, which, unlike a standard mirror, consists of a reflector with a large aperture and short focal length. This inexpensive modification allows for greater intensity for significantly lower cost. Despite its name, however, the plants using LFR or CLFR technology are not using actual Fresnel *lenses*, however, prototypes are in development. These lenses, made of glass, are theoretically much cheaper to produce than lenses.

LFRs are in use in several global applications including:

Puerto Errado 1 (PE 1)	1.4 MW	Spain	Novatec Solar
Puerto Errado 2 (PE 2)	30 MW	Spain	Novatec Biosol
Kimberlina Solar Thermal Energy Plant	5 MW	Bakersfield, CA	AREVA Solar

Top Countries Using Solar Thermal Power, Worldwide (GWth)

Rank	Country	2005	2006	2007	2008	2009	2010
1	People's Republic of China	55.5	67.9	84	105	101.5	117.6
–	European Union	11.2	13.5	15.5	20	22.8	25.3
2	United States	1.6	1.8	1.7	2	14.4	15.3
3	Germany	–	–	–	7.8	8.9	9.8
4	Turkey	5.7	6.6	7.1	7.5	8.4	9.3
5	Australia	1.2	1.3	1.2	1.3	5	5.8
6	Brazil	1.6	2.2	2.5	2.4	3.7	4.3
7	Japan	5	4.7	4.9	4.1	4.3	4
8	Austria	–	–	–	2.5	3	3.2
9	Greece	–	–	–	2.7	2.9	2.9
10	Israel	3.3	3.8	3.5	2.6	2.8	2.9
	World (GWth)	88	105	126	149	172	196

Solar Thermal Heating in EU (MWth)

Rank	Country	2008	2009	2010
1	Germany	7766	8896	9677
2	Greece	2708	2852	2859
3	Austria	2268	2518	2686
4	Italy	1124	1404	1870
5	Spain	988	1262	1475
6	France	1137	1371	1102
7	Switzerland	416	538	627
8	Cyprus	485	515	501
9	Portugal	223	345	471
10	Poland	256	357	459
11	United Kingdom	270	333	401
12	Denmark	293	331	368
13	Netherlands	254	285	313
14	Belgium	188	204	230
15	Sweden	202	217	227
16	Czech Republic	116	148	216
17	Slovenia	96	112	123
18	Hungary	18	58	105
19	Ireland	52	75	92
20	Slovakia	67	76	85

(Continued)

Solar Thermal Heating in EU (MWth) (*Continued*)

Rank	Country	2008	2009	2010
21	Bulgaria[a]	22	90	74
22	Romania[a]	66	80	73
23	Malta[a]	25	29	32
24	Finland[a]	18	19	23
25	Luxembourg[a]	16	19	22
26	Estonia[a]	1	2	2
27	Lithuania[a]	3	2	2
28	Latvia[a]	5	1	1
	EU27 + Sw (MWth)	19,083	22,137	24,114

[a]Estimation.

3 The New Culture of Environmentalism

Environmentalism is certainly nothing new. Philosophers, writers, and politicians have extolled the virtues of sustainability for hundreds of years. Over the last century in the United States, however, sustainability has typically been cast as a debate between environmental protection and economic growth. More recently, pioneering entrepreneurs have, in many ways, turned that debate on its head by demonstrating that environmentalism and economic growth are not mutually exclusive activities. This chapter will review the history of environmentalism and explain the key similarities and differences with the "green" movement currently underway in the United States and abroad.

One of the important premises of this book is that the factors driving the solar industry, such as globalization, capital markets, government intervention, public attitudes, and resource allocation, are interrelated and interconnected. Most of these factors are also clearly defined and routinely connected. Analyzing energy prices, for example, is a fairly straightforward endeavor.

Not so with public perception. This is for several reasons. First, it is a relatively recent phenomenon for pollsters to track perceptions in the United States and abroad. Second, terms of use change over time. For example, what was once called environmentalism may now be called sustainability, or clean, or green. Often times, terms with specific meanings are used interchangeably by people, further skewing our understanding. Carbon management is perhaps best understood as one form of sustainability or environmentalism, however, many use these terms to mean the same thing.

In order to reviewing the history of environmentalism in the United States and abroad and recognizing the limitations previously presented, this chapter will focus on two principal areas. The first is a review of public perception polling. The second is an increase penetration of so-called clean energy goods and services by the average consumer.

While these measures are perhaps more difficult to track, it is the opinion of this author that one of the significant differences between the previous solar boom and today is that society has undergone key changes, most notably a greater willingness to choose more environmentally friendly choices—even when more costly. This is particularly evident in the increased popularity of hybrid-electric vehicles. Perhaps as much as any other factor, the solar industry is boosted by consumers wanting to show the world that they have "gone solar."

The modern environmental movement in the United States is generally considered to have been catalyzed by Rachel Carson's *Silent Spring*.[1] This text, published

[1] Carson, R. (1962). *Silent spring*. Greenwich, CT: Fawcett.

Solar Energy Markets. DOI: http://dx.doi.org/10.1016/B978-0-12-397174-6.00003-9

in 1962, detailed the impacts of pesticides on birds and highlighted the damaging effects of chemicals on the environment. Carson's imagery captured the attention of the American populace and created a wellspring of support and advocacy for protecting the environment.

Many pundits still consider the environmental movement of the 1960s and 1970s to be nothing more than a philosophy pushed by a fringe group—often using the deriding term "tree hugger"—to minimize its potency. However, in the years following *Silent Spring* the United States experienced significant environmental policy changes, drastically altering the course of the nation's sustainability. The movement spawned the Environmental Protection Agency, the Clean Water Act, the Clean Air Act, and Superfund, just to name a few.

Much of the "early" environmental movement dealt with visible pollution—dirty air, unsafe water supplies, and hazardous waste. Little was known about carbon or climate change. As a result, the discussion about clean energy, and solar power in particular, was primarily focused on volatile organic compounds, sulfur oxides, and other emissions that were being attributed to acid rain, asthma, and cancer.

While most of the environmental movement continued to gain momentum through the 1980s and 1990s (despite attempts to stifle it by the Reagan Administration and Newt Gingrich's Contract with America), clean energy took a different path, one that has had significantly more highs and lows over time.

Global Trends

When attempting to measure changes in public perceptions, it is always difficult to select a starting point. This is specifically challenging when regarding sustainability environmentalism, two traits often cited by American Indian tribes is foundational, and lasting for tens of thousands of years. For our purposes, however, we will begin in 1972 with the United Nations Conference on the Human Environment in Stockholm.[2]

The 1972 Stockholm conference is widely regarded as the seminal moments of the modern environmental movement across the globe.[3] The basic premise of the 1972 UN Conference, which was repeated throughout the 1970s and 1980s, was that environmental protection was a luxury of wealthy nations.[4] This assumption, however, was tested by a massive study conducted by Gallup International in 1992.[5]

[2]Dunlap, R.E., 1994. International attitudes towards environment and development. In: Bergesen, H.O., Parmann, G. (Eds.), Green Globe Yearbook of International Cooperation on Environment and Development. Oxford University Press, Oxford, pp. 115–126.

[3]See Baylis, J., Smith, S., 2005. The Globalization of World Politics, 3rd ed. Oxford University Press, Oxford, pp. 454–455.

[4]Dunlap, at p. 115.

[5]*id.*

The 1992 survey, which was the most comprehensive global environmental survey to date, assessed residents of 24 nations that ranged in economic profiles and geographic regions. Interestingly, the survey showed that residents of all nations, regardless of economic attainment, view environmental degradation as a serious concern. Little has changed in this regard since 1992.

Around the same time, public perceptions and desires, as well as many policy debates, shifted from environmental protection versus growth and a new term "sustainable development" came into vogue. This way of thinking suggests that growth and sustainability are not in competition, nor mutually exclusive.

Perhaps the biggest change in global perceptions over the last 40 years is the public's focus away from traditional pollution and towards climate change. Like other environmental causes before it, the public is starting to recognize that clean energy does not necessarily translate into higher energy prices and strangled growth, but that in fact there are economic benefits to clean energy and greenhouse gas reduction in addition to environmental ones.

Recent research suggests that there is a strong correlation between media attention to global warming and the public's perception of it as a serious concern. According to a meta-analysis published in *Public Opinion Quarterly*,[6] "by September 1988, following record summer heat and a major upswing in media attention, awareness of the issue [in America] had spread to 58 percent of the public (from only 39% in 1986). More recent polls, after years of up and down media reporting, suggest that awareness of global warming has increased to between 88–91%."[7]

United States

Ample evidence suggests that "climate change" has become the major environmental concern in the United States, outpacing perennial front-runners of air pollution and water pollution.[8] Since 2003, the number of Americans rating climate change as a "serious" or "very serious" issue has risen dramatically. A *New York Times/CBS News Poll* from April 2007 suggests that 90% of the American public believe climate change to be a serious problem. In fact, over half (52%) of respondents noted that global warming was important to them personally.

Despite the heightened awareness of climate change, few Americans report that they understand the issue well (about 11% reported understanding global warming very well in 2007, basically unchanged from 1992 polls).[9] Global warming polls, taken nationally and across the globe do, however, indicate that more and more people

[6] Nisbet, M.C., Myers, T., 2007. Twenty years of public opinion about global warming. *Public Opin. Q.* 71 (3), 444–470.
[7] *id.*
[8] Washington Post Poll, 2007.
[9] *id.*

believe climate change to be real, to be connected to greenhouse gas emissions, and that climate change will pose a threat to them in their own lifetimes.[10]

Comparing US sentiment to other nations, particularly within the European Union (EU), is difficult. Surveys are not conducted uniformly meaning that the sample frames, questionnaires, and geographic distribution of responses are not consistent. Compounding these difficulties, sentiment across the 50 states in the United States and within each EU country can vary widely. Though the EU generally has higher knowledge about and understanding of climate change and renewable technologies, it is in the similarities that give pause to the solar industry.

According to a slew of recent polls, Europeans and Americans alike—despite being generally in favor of environmental protection—are not translating these sentiments into changes in everyday life. Significantly higher numbers of respondents are concerned about climate change than those who believe governments should spend money to combat the issue, and even fewer report making significant changes to their consumer patterns. This finding suggests two things: (1) that public perception is better used as an indicator for support of utility-scale projects or consumer incentives and (2) firms need to focus more intently on marketing the nonenvironmental benefits in addition to reduced greenhouse gas emissions.

While these statistics demonstrate a heightened awareness and concern regarding climate change vis-à-vis other environmental issues, the needle has moved more slowly relative to other national concerns. Since 2006, several polls have shown issues, such as the economy, national security, and healthcare. As recently as 2008, a Pew poll reported that Americans were more interested in developing new forms of energy than protecting the environment.

Comprehensive analysis of public perception is a critical component in forecasting future growth of the solar industry. It also provides important marketing cues for businesses and can predict policy changes. While polling suggests that the US public is slightly less focused on climate change and environmental protection, then their counterparts in Europe, many other indicators can and should be explored, especially considering that many public opinion polls on these matters contain social biases.

While public perception of climate change has a longer history and therefore allows for better longitudinal analysis, two recent polls specific to public opinion of solar offer important insights for the industry.

The first poll reviewed was conducted by Hart Research for the Solar Energy Industries Association (SEIA). The survey, fielded in early September 2012, was a representative sample of over 1200 registered voters and oversampled "swing voters." Overall, voters expressed a highly positive view of solar energy. In fact, 85% of voters have a very favorable (60%) or somewhat favorable (25%) view of solar energy.[11] As illustrated below, this puts solar atop all other tested energy sources.

[10] See Nisbet and Myers (2007).

[11] SEIA National Solar Poll, September 2012.

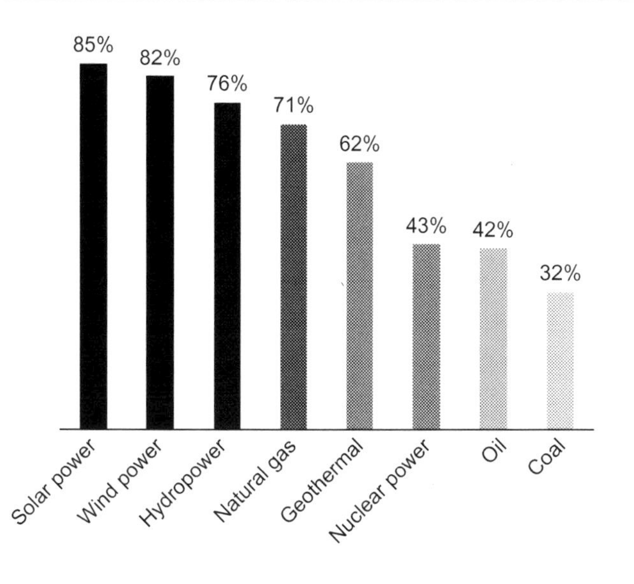

Interestingly for an election year with much attention paid to austerity, 64% of voters reported a desire for federal solar incentives and 70% of voters would like the federal government to do more (16% favor continuing its current policies and 14% prefer to see the government doing less).[12] Specifically, voters indicated that 78% of voters say the "federal government should provide tax credits and financial incentives to encourage the development and use of solar energy and only 22% say the federal government should not do this."[13]

Despite all of these positive news, respondents to that survey still believe that solar is too expensive (66%) to be practical for most consumers, while 54% believe that it is not practical for most parts of the country.[14] This last finding is particularly problematic because it is likely that the reasons for their answers are not rooted in the actual feasibility of solar. Kentucky, with few state incentives and very inexpensive electrical prices, is an example of a state that is less attractive for solar than Massachusetts, which has less solar resource but high electric prices and an attractive rebate program.

Public perception matters because at least one of the benefits of "going solar" is reducing carbon and other emissions in electrical production. Whether for gathering support for utility-scale operations, beneficial public policies, or making individual consumer choices, the fact that solar power produces clean electricity is becoming more and more attractive to US residents.

Of course, public perception does not exist in a vacuum, and clean energy has a lot more to offer than simply environmental improvement. Issues such as energy security and perhaps most importantly, cost, are critical. As a result, it is imperative to analyze changes in consumer spending choices and penetration of "clean and green" goods and services in the United States and abroad.

[12] *id.*
[13] *id.*
[14] *id.*

One important bellwether for solar markets can be found in hybrid-electric automobiles. Like traditional solar systems, hybrid automobiles typically have higher upfront costs and return on investment (ROI) for consumers is highly dependent on tax policy and energy prices. Hybrid vehicles also suffer from many of the same doubts by the public regarding their cost-saving potential. Because of these similarities, and due to the more mature and less regulated market in which they operate, reviewing sales of hybrid vehicles is a good and comprehensive place to start the analysis for solar product penetration.

Current estimates suggest that 2012 is a banner year for hybrid vehicles helping to push US auto sales back to pre-recession levels. One report estimates that hybrid vehicle sales are up to 67% since 2011, a year in which hybrid vehicle market share increased from 2.1% to 3%, according to industry analyst Alan Baum.[15] This success continues even as more fuel-efficient traditional gasoline engines invade the market.

Hybrid vehicle sales illustrate a new wrinkle in 2012—the uncoupling of sales figures with gasoline prices. For the first time ever, hybrid sales grew despite falling summer gasoline prices.[16] While this may merely be an indication that the US consumer believes that high gasoline prices are here to stay, there is clear evidence that changing consumer sentiment is at play.

In one recent survey, the Harris Group found that:

> *The adoption curve for hybrid vehicles appears to wane with age. While nearly one-third (32%) of those under 35 years of age are more interested in alternative vehicle choices than they were a year ago, the same can be said for only 15% of those over the age of 67. In fact, 11% of those 67 and older report that they are less interested compared to one year ago.*[17]

Perhaps even more importantly, the survey finds that a majority are basing their preferences on saving money (55%) while only about a quarter (26%) report environmental concerns.

Without clear and ongoing surveys of solar preferences, these patterns provide important context for the solar industry, and not only in the US Consumption patterns in the developing world (and China in particular) show low levels of penetration of efficient products, including hybrid-electric vehicles.

Europe

According to the Eurobarometer, nearly all (95%) Europeans feel that it is important to protect the environment, and, like their counterparts in the United States, a majority rank climate change as the most important environmental issue facing the world. So-called "old member" states generally view environmental policies and technologies

[15] http://www.jsonline.com/business/hybrid-car-sales-not-as-dependent-on-gas-prices-d16eo2f-165867746.html.

[16] *id.*

[17] http://www.harrisinteractive.com/NewsRoom/HarrisPolls/tabid/447/mid/1508/articleId/1059/ctl/ReadCustom%20Default/Default.aspx.

more favorably than "new member" states within the EU. Interestingly, however, there is growing evidence that policies are in fact leading the public rather than vice versa.

The most recent data concerning Europeans' perceptions are presented in a report by the EU from January 2013. Respondents from all 27 member nations were asked various questions related to air pollution, and specifically concerning the setting of energy policies and priorities.[18]

Fossil fuels fared quite poorly in the survey. Whether conventional (oil, gas, coal) or unconventional (shale gas, etc.), fewer than 10% of respondents across the EU were in favor. On the other hand, renewables enjoy very strong support with 70% in favor. This is particularly important given the recent economic downturn in Europe, as the data collected after austerity measures were put in place.[19]

Renewables are also much more popular than efficiency measures, which ranked favorably with only 28% of respondents. At the same time, "renewable energy sources are most mentioned by respondents in Portugal (82%), Austria, Spain, German, and Denmark (all 81%). In fact in only two countries are they mentioned by fewer than half of all respondents—Bulgaria (45%) and Romania (49%)."[20]

The report also includes a socio-demographic analysis. Somewhat surprisingly, there are no notable differences across groups regarding unconventional fossil fuels like shale gas with this option being mentioned by between 8% and 12% of each demographic. Also, men are more likely than women to prioritize nuclear energy (23% vs. 13%). Older respondents (55+) are the least likely to mention energy efficiency (24%) and renewable energy sources (65%), while each of those categories are mentioned more often as education level rises.[21]

China

Reliable measurements of public opinion in the developing world, as well as China, are less readily available than in other nations. However, a recent poll of 19 nations conducted by WorldPublicOpinion.org found that a majority of nations, 15 in total, has a citizenry that desires the government to do more to combat climate change. In China, a surprising 62% of respondents seek more action from their government. In fact, of the 19 nations surveyed, only Mexican residents placed climate change as a higher priority for their government.

China is also experiencing a significant youth movement that shows the early seeds of environmentalism. According to one recent study, 80% of young Chinese are concerned about global warming, however, this is not likely to change their consumption patterns. While 76% of the 2500 people surveyed said they did what they could to save energy, they still want a high-consumption lifestyle of a large house, car, and foreign travel.[22]

[18] http://ec.europa.eu/public_opinion/flash/fl_360_en.pdf.
[19] *id.*
[20] *id.* at p. 101.
[21] *id.* at pp. 102–103.
[22] http://news.xinhuanet.com/english/2007-08/20/content_6570747.htm.

Given these conflicting demands, Chinese expect government and technology to drive change. In the same survey, 78% reported that it was the government's responsibility to manage the environment. At the same time, the youth are flocking to hybrid-electric vehicles. This is another clear example that there is little appetite in China to pay more for clean technologies and less willingness to sacrifice standard of living.

There are no readily available, credible pubic opinion polls on Chinese solar perceptions. However, electric vehicle perceptions offer a critical window into the thinking of Chinese citizens' "green" consumption patterns.

According to surveys conducted by Boston Consulting Group (BCG), Chinese citizens view electric vehicles more favorably than in the United States or Europe. This may be the result of the government's promotion of electric vehicles, as well as a desire to be early technology adopters. That study showed that 91% of the Chinese surveyed were interested in electric vehicles, compared with 64% in the United States and 70% in Europe. Electric vehicles topped all other alternative fuel vehicles in China, while in Europe and the United States more respondents were interested in hybrid-electric vehicles as opposed to electric cars.[23]

A separate study by Neilson underscores these trends noting that "environmental awareness is rising along with higher oil prices in recent years, and more Chinese consumers are considering buying electric vehicles (EV), with a purchase intention of more than 50 percent. Compared to two years ago, consumers are willing to spend more for electric cars, and more than half (52%) of consumers are willing to pay a higher price for pure EVs, with a mean premium of 24,763 CNY, much higher than the 2009 level of 10,000 CNY. Consumers in Tier 1 cities such as Shanghai are willing to pay even more for EVs, 32,000 CNY on average."[24]

In fact, BCG found that 13% of Chinese were willing to pay more for an alternative fuel car even if they would never recoup the higher price paid, as opposed to just 6% in the United States and 9% in Europe. Perhaps even more telling, 53% of Chinese noted their willingness to pay more for a vehicle up front if they made it back over time, while only 44% of Europeans and 38% of Americans reported the same.[25]

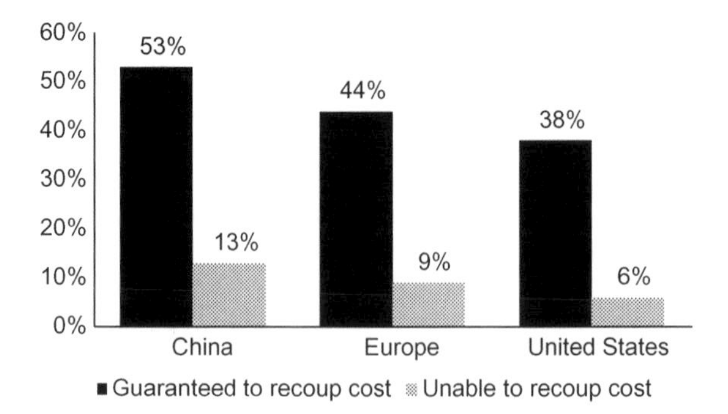

[23] BCG, powering autos to 2020.

[24] http://www.nielsen.com/us/en/newswire/2011/the-next-generation-of-chinese-car-buyers-are-looking-for-style.html.

[25] BCG, powering autos to 2020.

The new culture of environmentalism is driven by a movement of youth but is strongly supported by baby boomers in the United States, Europeans across demographics, in Asia, and across the developing world. Driven by heightened awareness of environmental issues, increased attention from national policy makers, and emphasis on new technology, consumers across the globe are racing to eco-friendly product options.

As the most popular renewable energy source across the globe and the one with the most consumer applications—from distributed electrical generation to domestic hot water heating—the solar industry is reaping and will continue to reap the benefits. However, cost remains critical. As can be seen across numerous clean technologies, a majority of consumers are concerned about higher upfront costs (particularly in the United States), especially when payback remains unclear.

At the same time, consumers seem positive but perhaps misinformed about the applicability and feasibility of solar power. Many would not consider a northern US state like Massachusetts to be a prominent solar market, however, with strong state policies, high electrical prices, third-party ownership, and lower component prices, Massachusetts is one of the fastest growing solar states and has the fastest time for ROI at only 4 years.

With greater awareness, continued installation growth, and ever-increasing environmental awareness, the solar industry is positioned for long-term, sustained growth all across the globe, despite potential short-term fluctuations.

4 Finance and Venture Capital

Venture capital and other high-risk startup funding are important components for the development and deployment of all new technologies. From both global and US perspectives, IT remains the largest recipient of venture capital, however, renewable energy technologies have made incredible strides over the past 10 years. In 2001, US companies received a modest $458 million invested by venture capital firms, only 1.2% of all Venture Capital (VC) funding in that year. In 2011, by contrast, nearly $6.6 billion were invested in US firms, a startling rise to over 23% of the total VC investment in that year.[1]

However, the path has not been even and 2012 was a down year for venture capital generally, and particularly sour for renewable energy and solar technologies. From a high of nearly $6.6 billion in 2011, US renewable energy venture capital dropped more than 20% to just over $5 billion, still a large historical increase, and down to only 19% of all venture capital invested.[2]

The future for US venture capital may not be terribly bright in the short term, as investors seek investments with fewer risks or shorter life cycles. For many, the sting of Solyndra, Evergreen, and A-123 systems are all too near and the lack of an Amazon-sized or Google-sized blockbuster remains elusive.

Like any high-tech industry, finance and venture capital play an important role in research, development, and deployment of technologies. In the solar sector, however, finance trends also impact construction and implementation of solar energy systems. While interrelated, these different finance options have specific and unique trends, pressure points, and mechanisms. This chapter will address R&D funding first, followed by a summary of global project capital forecasts.

Research, Development, and Deployment

The United States remains a global leader in innovation and its experience with the IT/software industry has significantly impacted its prioritization of capital flows vis-à-vis the rest of the world. The two main sources of private investment are venture capital and private equity firms. These firms invest in small, growth companies and buffer their high failure rates with explosive returns on the winners. This high-risk, high-reward system has worked extraordinarily well in the consumer goods market as it responds to a persistent drive towards innovative new products.

Renewable energy and solar in particular have not performed as well as their IT counterparts for many private finance firms, which are having a downward impact on

[1] CleanEdge 2012.
[2] CleanEdge 2013.

Solar Energy Markets. DOI: http://dx.doi.org/10.1016/B978-0-12-397174-6.00004-0

the innovation industry. A recent analysis by the NREL demonstrates the starkly different philosophies in the United States, Europe, and China.[3]

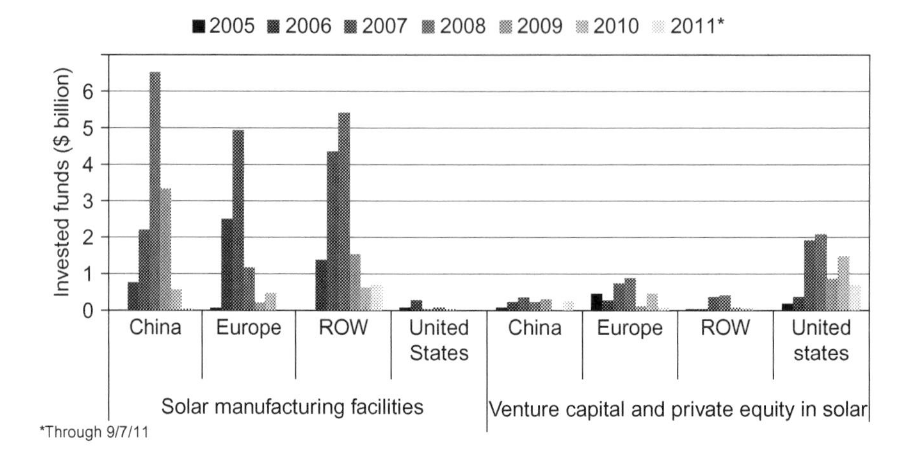

As can be seen in the figure above, the United States is significantly outpacing the rest of the world in R&D financing, but is way behind its competitors in investing in manufacturing (though declines in those nations suggest that the United States is coming closer to par).

Ultimately, the price curve and affordability of traditional photovoltaics are limiting the profit motive for investing in new technologies. With rapid declines in prices pushing the technology to near parity with fossil-fuel generated electricity, the justification for developing costly new alternatives, while more efficient, might take years to be cost-competitive. In other words, it is becoming increasingly harder to find companies willing to fund the quest to build a better mousetrap.

At the same time, however, market leaders in noncrystalline (e.g., thin film) technologies are ramping up their internal R&D spend, however, much of the focus is still on efficiency in the manufacturing process rather than development of new technologies.

2012 was the first year in a decade of declining global investment in renewables, down to $244 billion, or 12% lower than 2011 record ($279 billion invested).[4] This is a figure still higher than 2010 by nearly 8%.

While the decline is interesting in its own right, the difference between 2011 and 2012 is more starkly evident when reviewing the shift of energy investment by technology. For renewables overall, this was a marked shift from developed economies to developing ones. Specifically, investment in renewables was down 29% in 2012 in

[3]http://www.nrel.gov/director/pdfs/anu_public_lecture_10312011.pdf.

[4]Frankfurt School, UNEP Collaborating Centre, and Bloomberg New Energy Finance. Global Trends in Renewable Energy Investment 2013. Available from http://www.unep.org/pdf/GTR-UNEP-FS-BNEF2.pdf.

developed countries, while growing by 19% in developing ones.[5] This trend was especially stark in countries with policy uncertainty, such as the United States and Spain.[6]

Much of this shift was driven by China. After generally mirroring the United States in renewable energy investments over the past several years, in 2012 China asserted its dominance, increasing its overall investments by 22%—and primarily in solar activities. This shift to China with a further shift from production to installation as well as declining PV prices led to an overall increase in solar capacity despite an 11% global decline in solar investment in 2012.[7]

Private Investment Trends

Private investment trends are important to understand the future of the industry because capital tends to flow to the areas of greatest impact. It is important to keep in mind, however, that private investments only show once slice of the total, as many managed economies like China have seen enormous public investments in the solar industry over the last decade. These investments are covered following the data on private investments.

While understanding the various technologies is important to forecasting trends, even more important is analysis by stage. These stages include the following categories:

- *Concept*: This phase refers when a company's product, service, or technology is still in a planning or design phase.
- *Product development*: This phase refers to when a company is actively pursuing development/prototyping of alpha, beta, pilot, or commercial offering of its product, service, or technology but has not yet achieved customer traction.
- *Shipping product/pilot*: In this phase, a company has moved from product, service, technology design and development stage and has demonstrated initial customer traction or pilot deployments with strategic partners.
- *Wide commercial availability*: In the final phase, a company's product, service, or technology is widely available to the target market via commercial channels.[8]

Global Investments

Overall, global solar investment from 2007 to 2012 has been record breaking. Over that period, there have been 1283 private investment deals with a total of value of $44,661,340,063. These have been most frequently funded to firms that already have commercially available products across the globe, as seen in the figure below.

[5] *id.* at p. 11.
[6] *id.*
[7] *id.*
[8] Cleantech Group's i3 Platform.

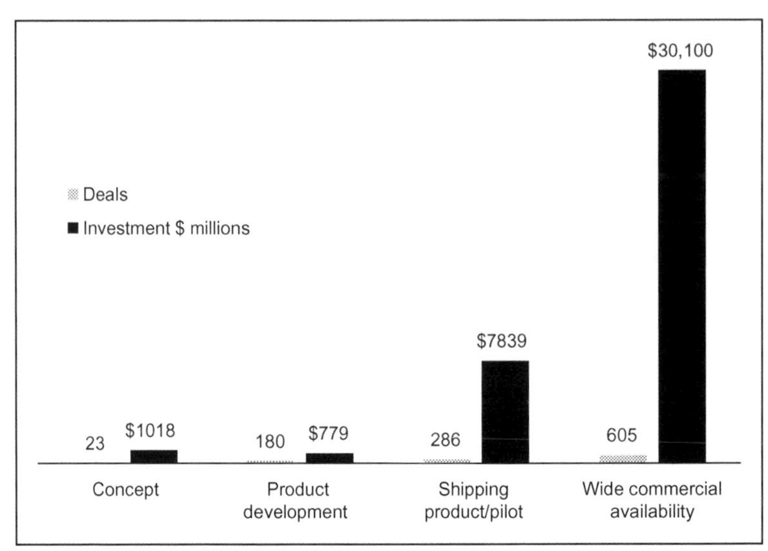

Global private investment, 2007–2012, by type, in millions.

Most of the investment during the period occurred during 2011 and 2012. As verified from other sources and trends, the 2012 declines do not look as precipitous when compared to previous years.

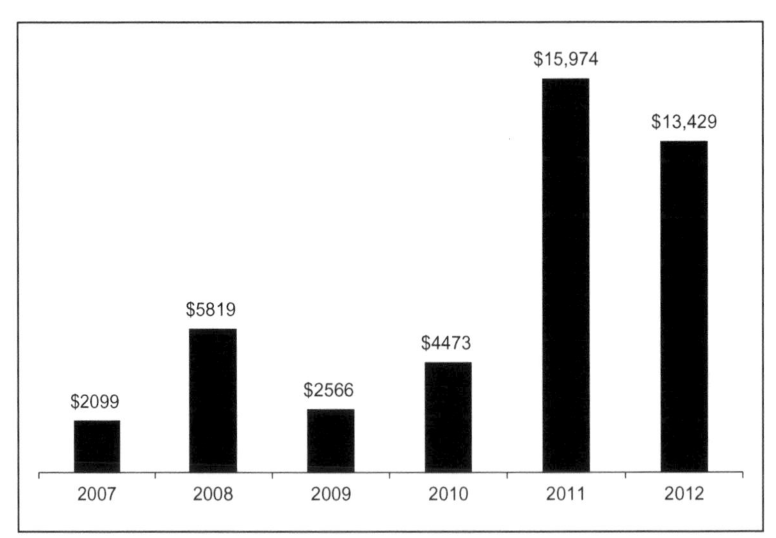

Global private investment, in millions.

From a technology standpoint, investments have been uneven. Concentrated solar power technologies, while up by about 350% from the 2010 level in 2012, dipped by about half from their peak in 2011.

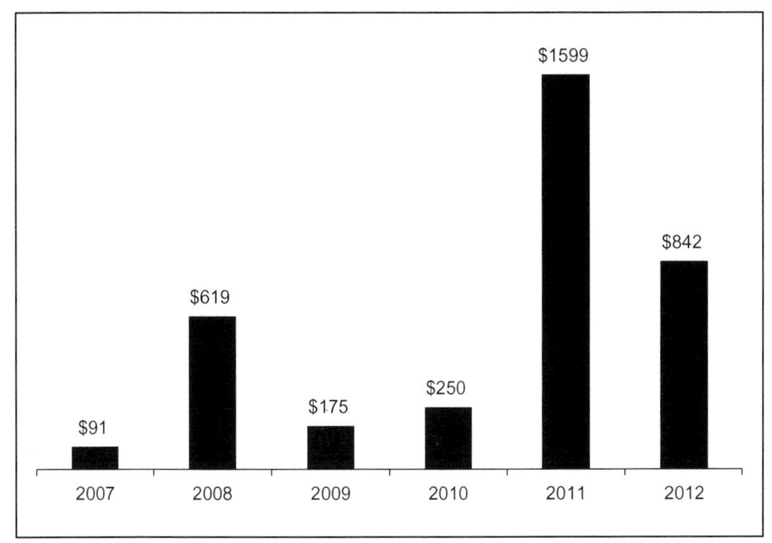

Global CSP private investment, in millions.

Declines in photovoltaics are much less dramatic, as seen in the figure below. This is likely due to continued price declines in PV products and greater cost competitiveness of the technology's deployment.

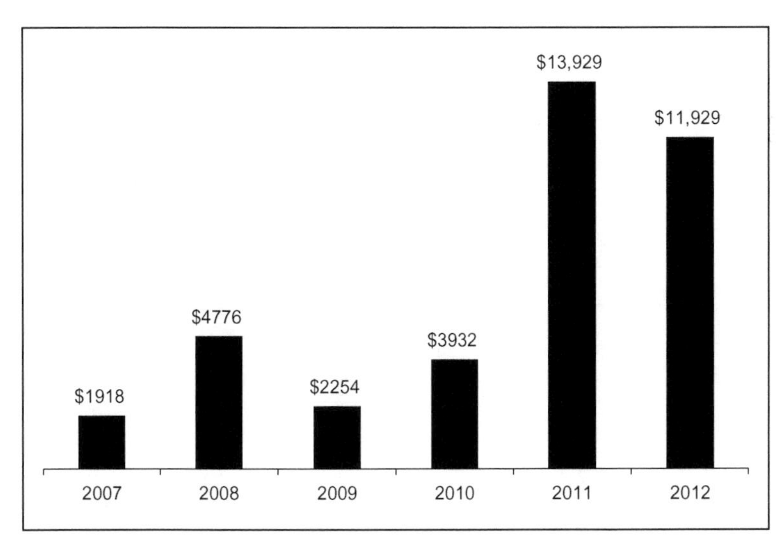

Global PV private investment, in millions.

Comparative Investments

The charts below illustrate the shifting trend to developing nations. As can be seen in the charts below, the United States and Europe are in decline, while Asia

Pacific, South America, and the Middle East are growing rapidly (all chart figures in millions).

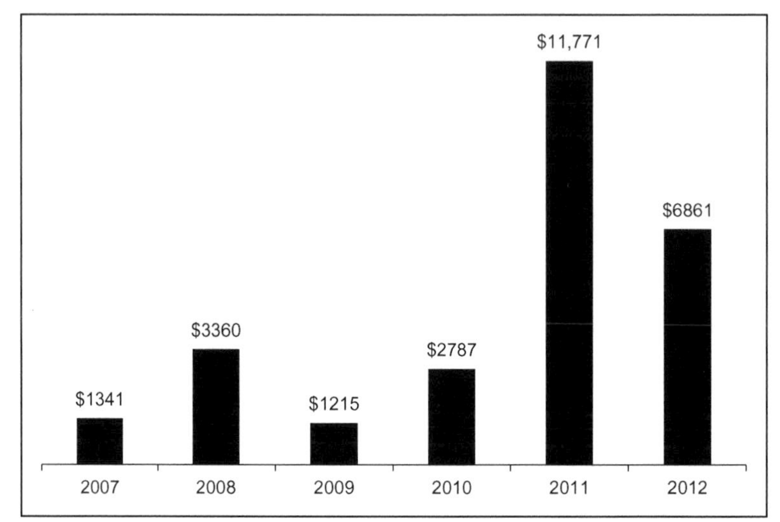

United States.

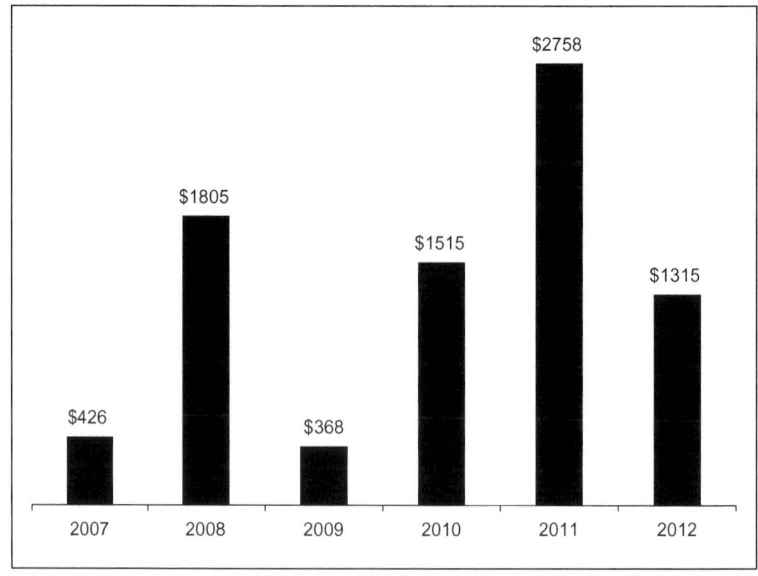

Europe and Israel.

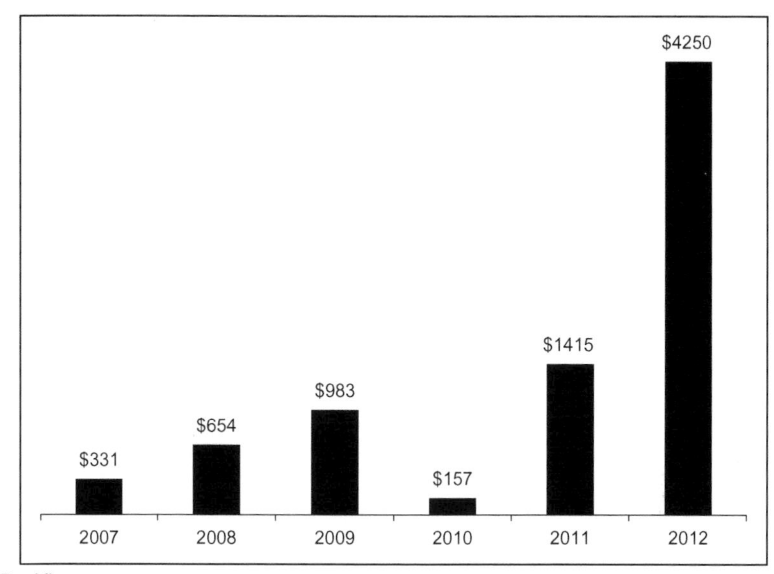

Asia Pacific.

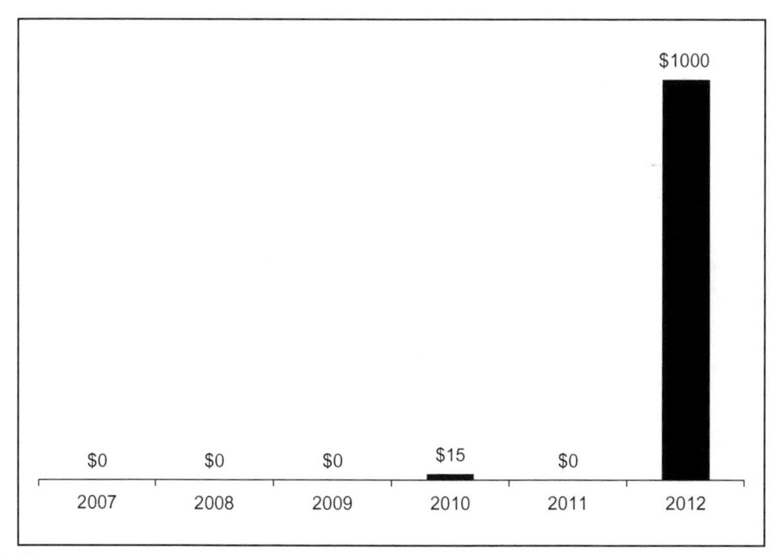

Middle East.

It is important to note again that these data include only private investment, which is why Asia Pacific's figures are significantly lower. In 2011–2012, it is clear that the United States is the clear leader in attracting private investment into the solar industry (despite China's overall lead when public investments are included).

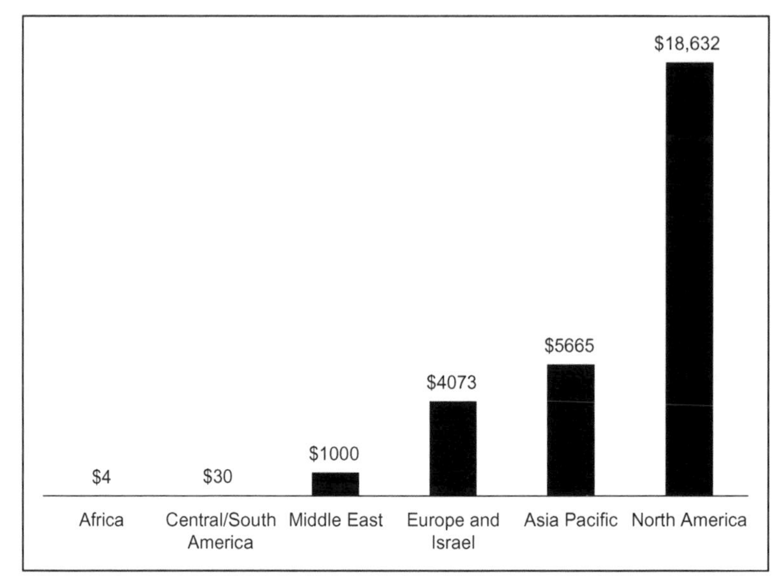

Private solar investment, 2007–2012, in millions.

United States

Solar firms in the United States received nearly $27 billion ($26,716,373,939) in private financing between 2007 and 2012 in over 700 investment deals. Most of this funding came for later-stage firms, as seen below.

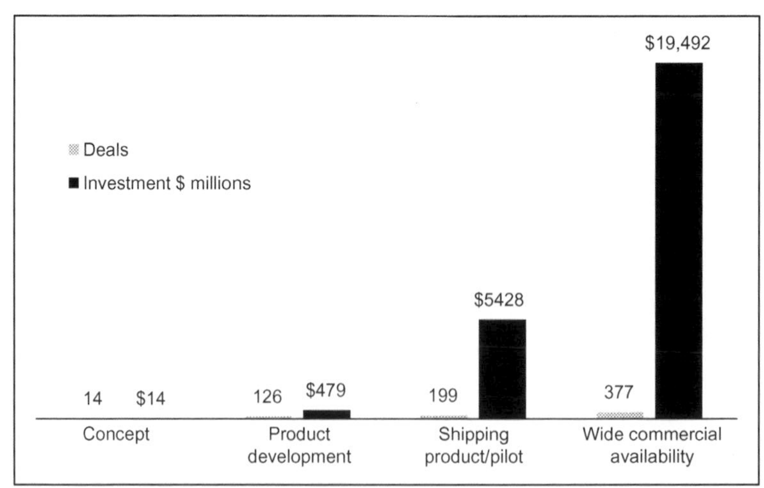

US solar investment, 2007–2012, by type, in millions.

Private solar investment spiked in 2011 followed by sharp decline in 2012. Despite the decline, it is still significantly higher than in 2010.

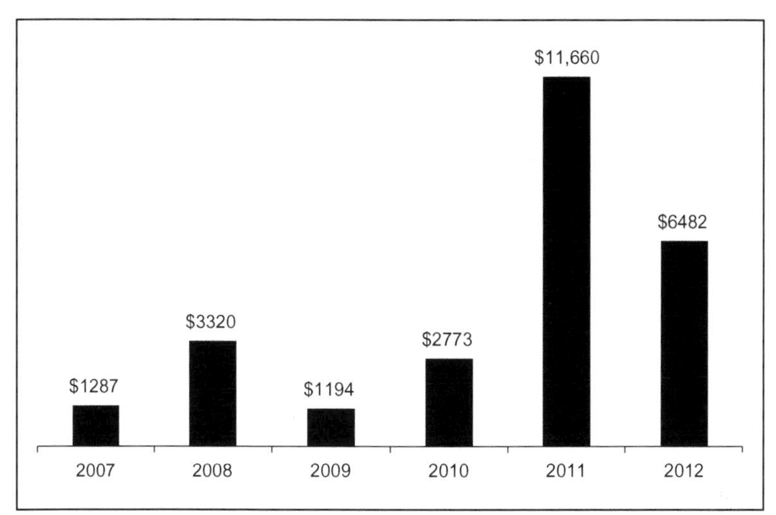

US private solar investment, in millions.

Germany

Germany has a much more volatile private investment market without clear trend lines. This makes for difficult planning of firms and is explained by recent economic turmoil and policy uncertainty.

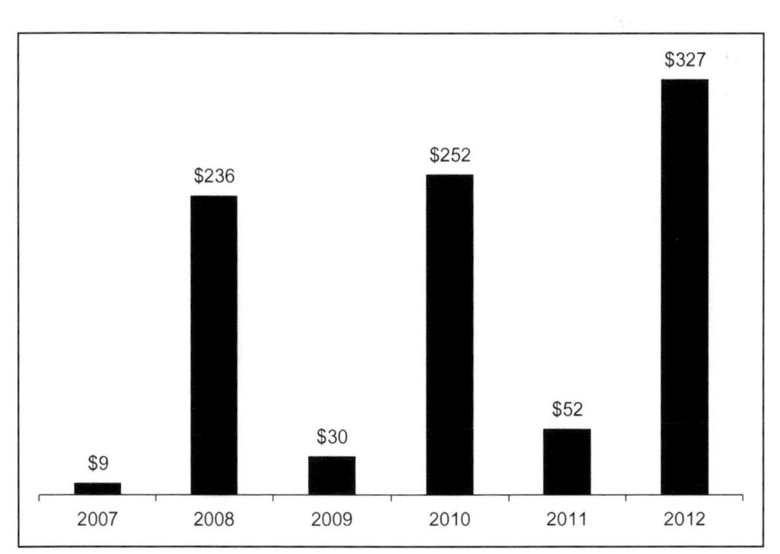

German private solar investment, in millions.

As with the United States, most of Germany's private investments go towards late-stage firms.

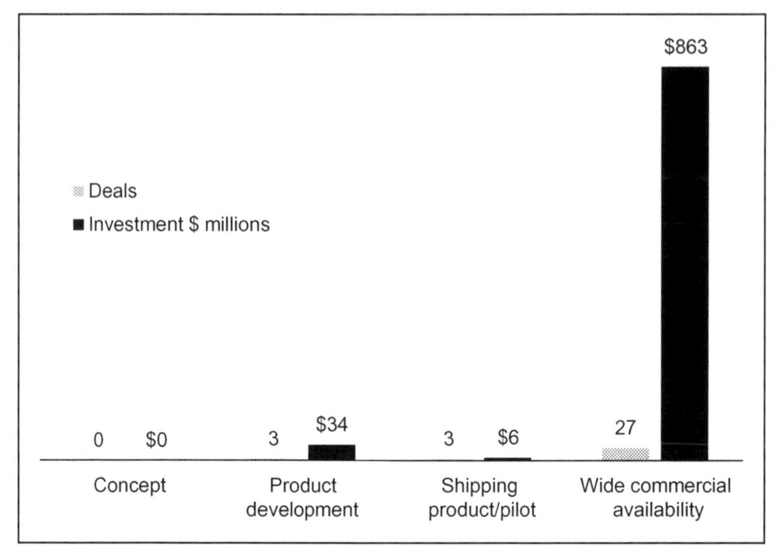

German private solar investment, 2007–2012, by type, in millions.

Spain

Spain's financial woes and policy choices have set the country back significantly in the solar sector, as illustrated by declining investments (2012 was similar to 2008 and 2013 looks even worse).

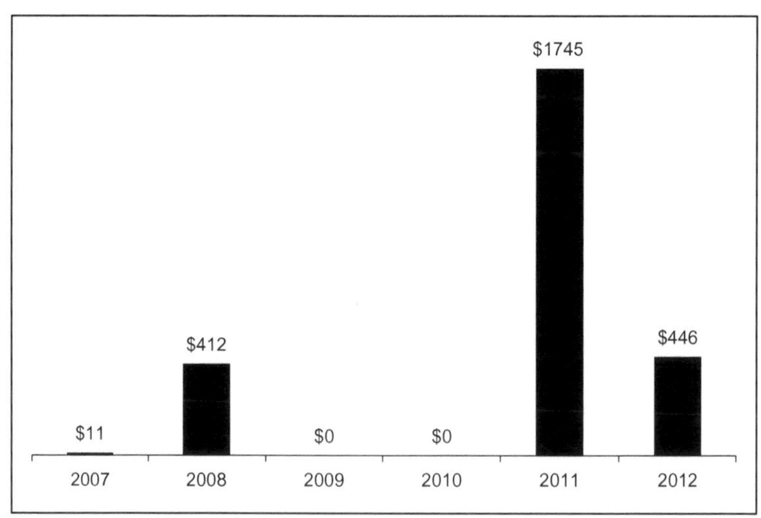

Spanish private solar investment, in millions.

Nearly all of the investment capital in Spain from 2007 to 2012 have gone to later-stage firms, consistent with the global figures and that of other developed countries.

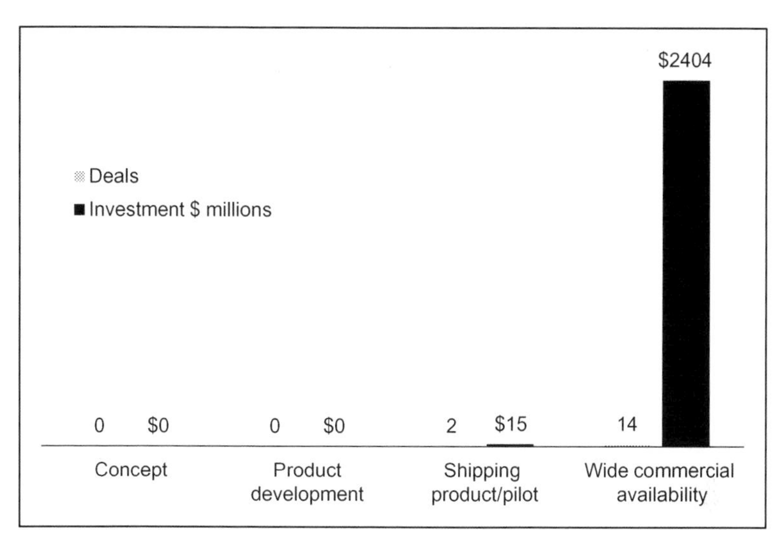

Spanish private solar investment, 2007–2012, by type, in millions.

Italy

Italy's private investment, according to i3 Cleantech Group data, has fallen off a cliff since highs of 2010, though it has never been a significant component, with only six deals over the period.

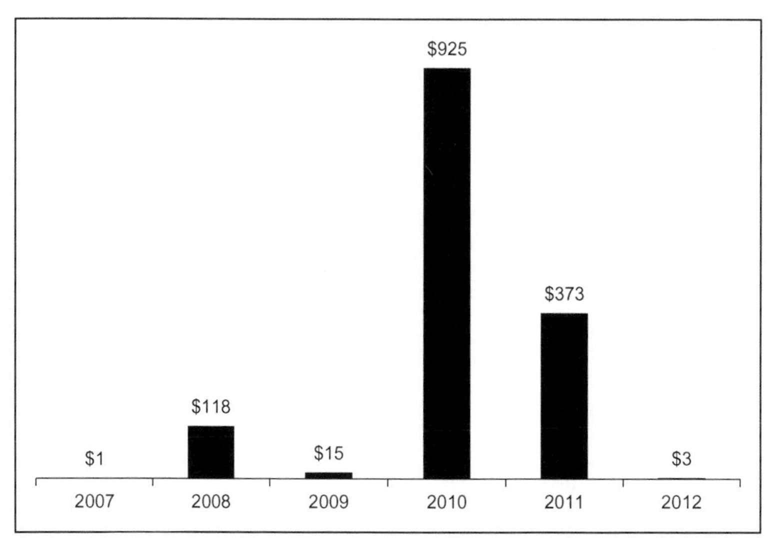

Italian private solar investment, in millions.

China's investment is traditionally public sector driven, however, there were 73 deals and about $6 billion in private capital invested between 2007 and 2012 with the majority coming in 2011 and 2012.

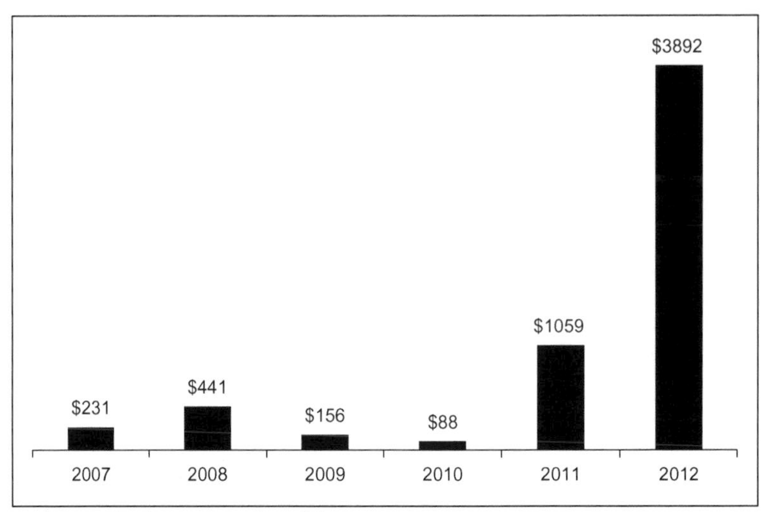

Chinese private solar investment, in millions.

Chinese private investments are still later-stage focused, but with a greater share going to stage 3 companies than in the developed world.

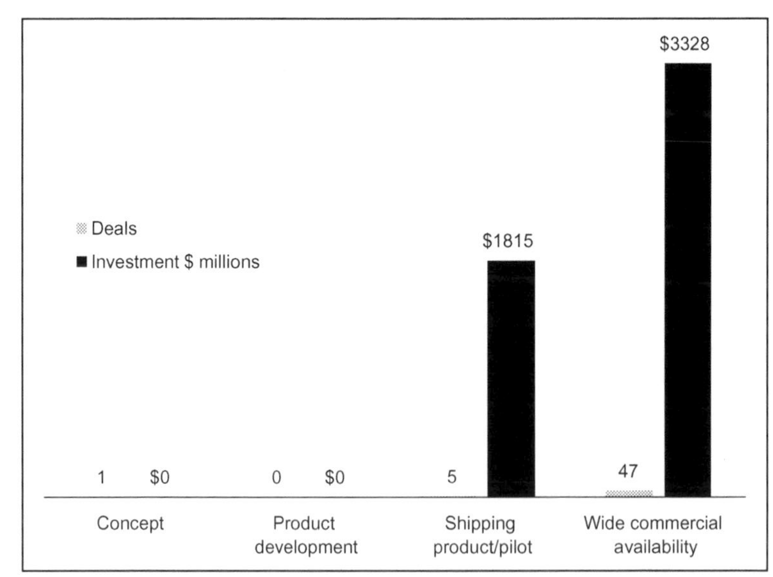

Chinese private investments, 2007–2012, by type, in millions.

India

India is making a strong push into the solar market, strongly rebounding after the global recession to approach 1.4 billion invested from 2007 to 2012.

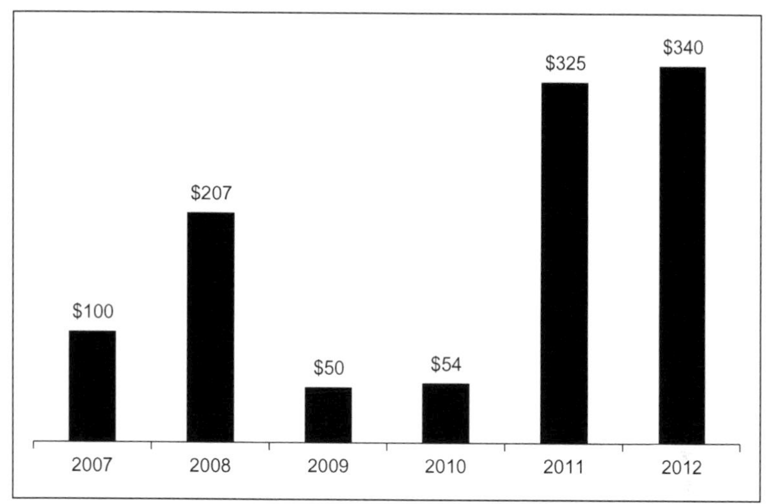

Indian private solar investment, in millions.

India, similar to other developing nations, has a wider spectrum of solar private investments with representation across all stages of companies.

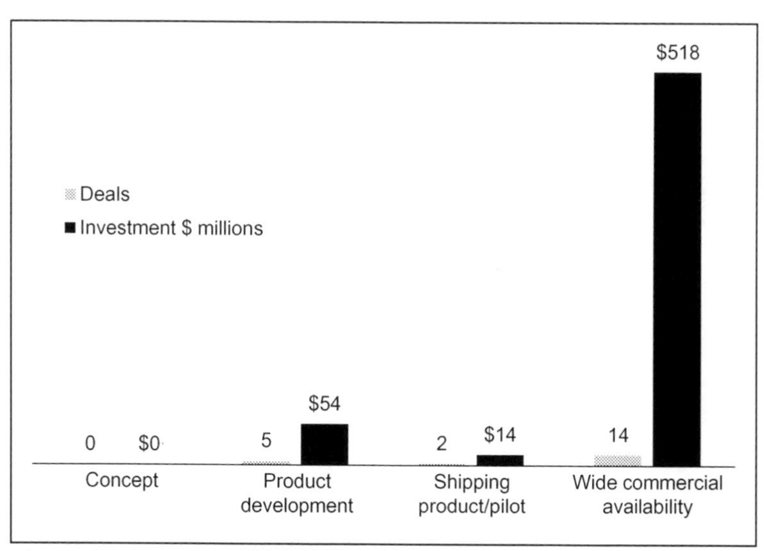

Indian private solar investment, 2007–2012, by type, in millions.

Despite these impressive trends in investment over the last several years, generators continue to ramp up fossil assets. In addition, share prices of renewable energy firms, in large part due to major declines in panel production profitability, fell sharply in 2012.[9]

[9] http://www.unep.org/pdf/GTR-UNEP-FS-BNEF2.pdf, at p. 11.

On the other hand, small-scale investments are up significantly, while asset finance of larger projects declined in 2012.[10]

Public Sector Spending

It is difficult to parse the differences between public sector spending and supportive policies; between reduced tax receipts (currently the stimulus measure preferred in the United States) and direct cash incentive (in favor in China). This brief section addresses several of the most important points for public sector trends across the globe.

- Europe, once the primary driver of solar markets, is embroiled in austerity measures. Little direct support for R&D or installation should be expected.
- In the United States, despite several high-profile bankruptcies, public investment in solar R&D continues to rise. In 2012, the $629 million invested in R&D by the federal government is nearly double the 2009 figure of $343 million. The *Energy Innovation Tracker* estimates that the most likely scenario is for this number to swell to $770 million by 2015.
- China is shifting its subsidies, however, its managed economy makes it impossible to compare on an apples-to-apples basis. Overall, China invests about $6.4 billion in wind and solar per year, a number sure to rise as installation becomes a major focus of the government to reduce environmental damage.
- New entrants, such as the United Arab Emirates and Saudi Arabia, are reinvesting oil revenues into solar to stem the trend of becoming net oil importers by 2030.

Overall, inexpensive capital, lower prices, and supportive policies have shifted investment starkly away from production and also from research and development toward installation. With the market economics (see Chapter 9) remaining favorable over the short- and medium term, it is a good bet to think that this shift will remain in place over time, even if its momentum slows. Equilibrium of supply and demand is critical for overall success in the sector, however until then, investments for developing a better mousetrap or producing more of them are not as favorable as development projects and this is unlikely to change in the near term.

[10] *id.*

5 Global Solar Policy

Government policies are important drivers of solar markets, impacting both the supply of and demand for solar energy. There are three primary policy objectives that most significantly impact the solar industry. The first of these, energy policy, impacts the supply side of the industry by generating incentives and mandates that serve to increase overall supply and increase the share of solar power in overall energy production. Examples of such energy policies include renewable portfolio standards, consumer tax incentives, and feed-in tariffs (FiTs).

Economic development is the second key policy objective that impacts the solar industry. While economic development policy also impacts supply, it does so with different mechanisms. Economic development policies spur innovation and production of clean energy goods and services. This industry creation objective is different from energy policy because it is not directly related to domestic energy mix. Rather, it is focused on creation of an industry that can be quite export driven.

The third policy driver is demand based, which manipulates energy markets and guarantees demand. Such policies, which are rare in the United States, ensure consistent demand for solar power. This chapter will explore all three types of policies with a particular focus on Spain and Germany in Europe, United States and several of its key regions, and China and the developing world.

Energy Policy

Energy policy is of obvious importance to solar markets. States and nations typically endeavor to develop clear and stable policies that weigh issues of cost, security, environmental impacts, and source diversity. Unlike many of its global competitors, however, the United States does not have a coherent nor comprehensive energy policy. This section will review energy policies in the EU with a specific focus on Germany and Spain, as well as current policy in China and specific attempts in the United States and its major states to plan for the future.

EU Policy Framework

Despite numerous attempts in the past, a consistent and common energy policy in the EU has been elusive. After Russia prevented the flow of gas into the Ukraine in 2006, Europe reacted with a new plan focused on energy security; however, it was not until the 2007 Treaty of Lisbon and the passage of the Energy 2010 initiative that Europe maintained a consistent energy policy.

Solar Energy Markets. DOI: http://dx.doi.org/10.1016/B978-0-12-397174-6.00005-2

The energy plan for Europe, or EPE, has three primary objectives: (1) to promote renewable energy, (2) to secure foreign energy supply, and (3) to encourage research and development of energy technologies. The consistency provided in a common plan allows for better ability to plan and negotiate as a unified bloc.

European energy policy has its basis in Article 194 of the Lisbon Treaty. The treaty is the framework for the common goals of the EU and establishes a commission and division focused on energy. This energy division is "part of a re-compartmentalisation scheme meant to address inconsistencies, overlap and gaps in the Commission's governance of energy."[1] This presents an interesting paradigm, where for issues of international energy importance, the commissioner for energy has primary jurisdiction, as opposed to the foreign minister.[2]

Despite this arrangement for matters within the EU, energy projects or policies that include countries outside of the EU are handled through a hybridized leadership team.[3] This arrangement was in clear display during the negotiation of the 2009 EU–US Energy Council, which was led by the foreign minister. According to *Braun*, this arrangement is further exemplified in the development of the Energy 2020 strategy,[4] noting that "EU policy will pay particular attention to safety and security of oil, natural gas pipelines and related production and transport infrastructure by combining energy policy and CFSP instruments."[5]

Navigation of these realities can be tricky. From a policy perspective, it is clear that member sovereignty and shared governance present specific challenges to future EU policy (an issue certainly not limited to energy policy). Clarity in this regard is critical for coherent policy and proper messaging to energy business concerns. Whether at the cabinet level or in the European parliament, progress will only be achieved if the global market understands the delineations of power, yet such clarity seems unlikely.

EU Supply-Side Policies

Proper understanding of this framework is important for determining the value of EU directives and policies related to solar. The 2007 EU Energy Roadmap outlines specific renewable energy targets for each member state with an overall requirement of 20% renewable energy mix by 2020. The 2010 Europe 2020 plan mandates national energy plans for each member, but as illustrated previously in this chapter, sovereign concerns, intra-governmental conflicts, and general economic and debt concerns in Europe give at least some pause despite many countries' expectations of exceeding their 2020 goals.

Europe's Renewable Portfolio Standard

The EU's Directive on Electricity Production from Renewable Energy Sources, or RES Directive, was passed in 2001 and updated in 2007. This directive, while not

[1] Braun, J., 2011. EU Energy Policy under the Treaty of Lisbon rules. EPIN 31, citing Andoura et al. (2010).
[2] *id.*, at p. 5.
[3] *id.*
[4] *id.* at p. 6.
[5] See European Commission (2010).

specifically binding, sets a target of 33% renewable electrical production and 20% renewable energy use by 2020. To meet these goals, the EU has established a series of national targets for individual member states, but nations are free to enact higher targets. The program is overseen by the European Commission, which has the power to set mandatory targets if goals are missed. As with most renewable portfolio standards, targets for specific technologies (e.g., solar wind) are not included.

Europe's Consumer Incentives

Consumer incentives are critical to growth of residential PV and water heater installations in the United States, driving down the upfront costs and reducing payback time. Direct consumer incentives, such as tax credits and subsidized financing, are less prevalent in Europe and there is no EU-wide program tied directly to purchasers of solar systems. A strong FiT policy is the primary incentive driver in European nations.

The EU is obviously a loose conglomeration of member states without sufficient power to directly incentivize or spur development of specific projects. While the goals and mandates set but the European Commission are clearly important to Europe's renewable future, the specifics are mostly driven at the national level. Three markets are critical to the future of Europe's solar industry: Germany, Spain, and Italy.

Germany

In 1991, Germany passed the world's first FiT law requiring utilities to pay for surplus solar power at a competitive rate. This law propelled Germany into a position of global leadership in renewable energy policy that it has only cemented over time, particularly with the passage of renewable energy targets in 1997 and the Renewable Energy Sources Act of 2000 (EEG).

Germany is the world leader in renewable energy policy, thanks in large part to the passage of the EEG. The EEG is an aggressive mix of incentives and mandates that focus on energy efficiency and renewable energy production. It is based on the premise that large initial investments yield economies of scale over time, driving down the overall price of energy and renewable energy in particular.

The EEG is perhaps the most comprehensive and aggressive approach to promote renewable energy, using incentives and mandates to dramatically alter the country's energy future and have served as a model for the rest of the world. This analysis reviews the supply-side impacts of German policy with an eye to future trends that may indicate storm clouds brewing for German markets.

Germany's overall climate and energy policies have ambitious goals, which easily surpassed its Kyoto Protocol targets. For 2020, German policy requires a 40% reduction in greenhouse gas emissions (as compared to 1990 levels). This will be accomplished by increased renewable production and greater energy efficiency measures. The most important of these goals for the solar industry are the electrical generation targets (for PV) and heat supply targets (for solar thermal).

German Supply-Side Policies

Electrical generation targets are critical for the future of the German solar industry, however, PV remains the smallest of all major renewable energies in Germany, making up about 5% of total electrical production. However, the permit requirements, potential for distributed generation, and economics of solar power make it easier and faster to deploy. As a result, time-sensitive targets tend to benefit solar over other technologies.

Renewable Portfolio Standard

German policy mandates that renewables provide 35% of the electrical mix by 2020 and a whopping 80% by 2050. For overall energy usage, the targets set are 18% by 2020, 30% by 2030, and 60% by 2050. To meet these supply-side goals, Germany has adopted a mix of supply-side policies, economic development measures to grow the industry, and demand-side requirements to force consumption of renewable power.

Feed-in Tariff

One of the key components of the EEG and the foundation of Germany's success in PV installation is a guaranteed FiT. An FiT is essentially payment for energy produced and distributed into the grid. Almost all renewable energy policies include FiTs to ensure that businesses and homeowners receive compensation for the excess power that they produce for the grid. Germany goes a step further, however, by guaranteeing a fixed price for 20 years. Not only does this provide an important incentive by reducing the overall ROI time scale but also provides long-term stability. This stability is critical to the development of renewable energy industries and particularly for the solar industry.

Germany is in the third iteration of its FiT scheme. The initial FiT policy was enacted to promote expansion of PV and other renewable installations at a rapid pace in order to promote domestic energy production. This scheme was in place from 2000 to 2009, a period when PV modules were quite expensive by today's standards and electricity produced required significant subsidies to be cost competitive.

With the virtually unchanged and generous FiT combined with the precipitous drop in panel prices in 2009, an investment and installation boom followed. The result of this tremendous added capacity was a strain on the financial scheme, and a potential for a boom–bust cycle, prompting a response from the German government. The changes to the policy in 2009 were flexible, resetting the FiT every 6 months based on capacity added. However, it was widely reported that the system did not work well, prompting additional changes with significant consequences to the German PV market.

By 2010, it became clear that renewable energy production, and particularly solar electricity, was approaching grid parity with other electric sources and that government policy was needed more to manage the energy portfolio and foster technical innovation. As a result, significant digressions in the FiT were realized. In addition to the regular digressions, a one-time cut of 13% in July 2010 and a 3% in October 2010 were made as well as a 13% breathing cap for 2011 to adapt to price fluctuations in the previous year.

Despite these cuts, further digressions were made in 2012, continuing a trend of reduction that is illustrated in the table below.

Peak Power Dependent FiT for Solar Electricity in €-ct/kWh

Type		2004	2005	2006	2007	2008	2009	2010	July 2010	October 2010	2011	2012
Rooftop mounted	Up to 30 kW	57.4	54.53	51.80	49.21	46.75	43.01	39.14	34.05	33.03	28.74	24.43
	Between 30 kW and 100 kW	54.6	51.87	49.28	46.82	44.48	40.91	37.23	32.39	31.42	27.34	23.24
	Above 100 kW	54.0	51.30	48.74	46.30	43.99	39.58	35.23	30.65	29.73	25.87	21.99
	Above 1000 kW	54.0	51.30	48.74	46.30	43.99	33.00	29.37	25.55	24.79	21.57	18.33
Ground mounted	Contaminated grounds	45.7	43.4	40.6	37.96	35.49	31.94	28.43	26.16	25.37	22.07	18.76
	Agricultural fields	45.7	43.4	40.6	37.96	35.49	31.94	28.43	–	–	–	–
	Other	45.7	43.4	40.6	37.96	35.49	31.94	28.43	25.02	24.26	21.11	17.94

Consumer Incentives

While the FiT has dropped dramatically over time in Germany, the Market Incentive Program, or MAP, remains strong. The MAP, which is administered by the Federal Ministry of Environment, provides direct subsidies and subsidized financing for various renewable energy projects, including solar heating technologies.

Current solar-related incentives include a base incentive and an innovation bonus based on the size of the collector. This funding, while crucial for the development of these markets, is in jeopardy given recent austerity measures in Germany. A brief summary is included below as developed by Intelligent Energy Europe.[6]

Example Measures	Base Support (Existing Buildings)	Max. Accumulated Bonus (Existing Buildings)
Solar collectors (warm water/ cooling demand) <40 m^2	0 €/m^2	180 €/m^2 innovation bonus
Solar collectors (warm water and heating) <40 m^2	90 €/m^2	180 €/m^2 innovation bonus or one of the following options: €500 combination bonus or 0.5 × base support efficiency bonus and/or €50 solar pump bonus

German Economic Development Strategies

While a nation's energy policy is clearly important to its solar industry, such policies tend to have an outsized impact on the installation and sales sectors. Given the global marketplace for goods, global demand is a more critical driver for the manufacturing and R&D sectors.

Germany is considered a world leader in manufacturing, bucking the commonly held belief that developed nations cannot compete in the global production of goods. The advanced manufacturing sector in particular has grown in Germany post-recession, however, continued sluggish global growth is producing potential storm clouds over the manufacturing sector in general, and coupled with fierce competition from China, an even greater concern for solar manufacturers in Germany.

Recessionary Impacts

Germany was battered by the Great Global Recession of 2009–2010, but as illustrated in the figure below, it rebounded quickly, based on the strength of its manufacturing core.

[6]Teckenburg, E. et al., 2011. Renewable energy policy country profiles. Intelligent Energy Europe, p. 119.

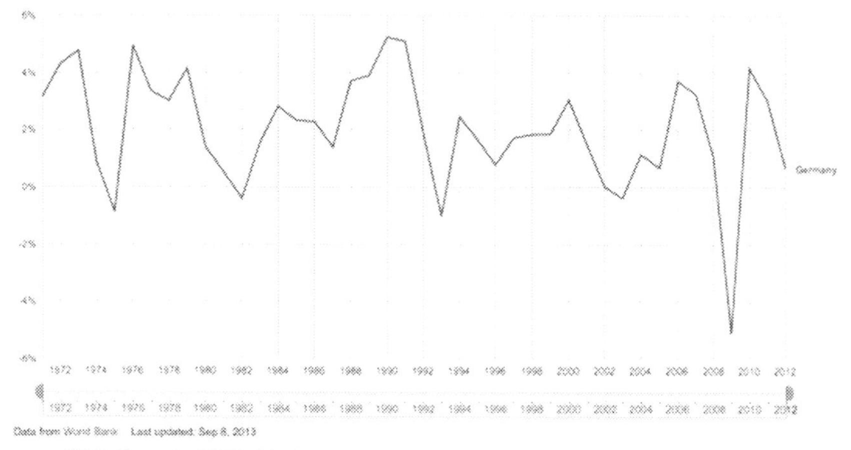

German GDP Growth: 1972–2012.
Source: Google Public Data using World Bank Figure.

However, storm clouds are brewing and the Euro Crisis and its related fiscal challenges seem to be drawing most of Europe, Germany included, into another recession. The Organization of Economic Cooperation for Development (OECD) predicted that Germany slid into recession in the second half of 2012 and with 0.5% and 0.8% contraction in Q3 and Q4, respectively.

OECD further predicts that the German success of 2008–2009 is unsustainable suggesting "a return to lower growth rates from the strong prior upswing was to be expected from a cyclical perspective as potential growth remains weak. This downswing is exacerbated by the substantial deterioration of world trade growth and a loss of confidence due to the euro area debt crisis."[7]

Even during the boom times, Germany rapidly lost share of the solar export market, falling from 77% share in 2004 to 31% in 2009. However, the steady 20% expected growth in solar year over year should strongly benefit this global leader in panel production and open newer markets for solar thermal technologies.

Growth Versus Austerity

The major trend to follow for the German solar export market is whether the country takes an austerity or pro-growth policy tack—and whether there is actually an opportunity (as some suggest) of achieving both goals. As of January 2013, it appears that Germany will not pursue an aggressive growth agenda, however, this is a trend that should be tracked closely by solar firms to determine the likelihood of additional stimulus.

[7]OECD Economic Surveys, Germany, February 2012. Available from http://www.keepeek.com/Digital-Asset-Management/oecd/economics/oecd-economic-surveys-germany-2012_eco_surveys-deu-2012-en.

Manufacturing and R&D Supportive Policies

German manufacturers have long benefited from a cogent domestic policy that subsidizes exporters. The primary mechanism for this subsidy is a value-added tax (VAT), which has been set at 19% or 2% higher than the EU average. The VAT has specific importance to the solar industry, as so much of the German product is exported to America, which has no VAT. As a result, exporting manufacturers receive a VAT refund and are not subject to a VAT in other markets. American manufacturers, on the other hand, are charged a 19% VAT for any products sold in Germany. This basic tax structure has significant impacts on the competitiveness of German-made solar products vis-à-vis its global competitors and also raises important questions about how multinationals should make citing decisions for their operational bases.

Despite the favorable export policies enacted by the German government, the solar manufacturing industry has lost significant market share to Chinese and other global competitors. The competition has dramatically decreased prices and substantially eroded profit margins. At the same time, the low current cost of crystalline PV panels has depressed research and development of new technology, given the cost competitiveness of current products.

Ample funding of R&D projects exists, despite recent budget cuts in Germany. The largest and most important of these schemes is direct project funding or DPF (for solar, it is under the program for Environment and Energy). This program provides funding for research institutions, private companies (or partnerships among them) to develop new and innovative technologies over a longtime horizon.

According to a 2012 forecast by *Batelle*, overall German R&D expanded in 2010–2012 in real terms and is poised to continue to grow.[8] Of course, as austerity measures are implemented, public research funding is likely to decrease. Private investment in research has become pivotal and its importance will grow over time.

In addition to the direct funding programs, the Federal Ministry of Education and Research (BMBF) notes the following important indirect funding mechanisms to support the German economy:

- German Federation of Industrial Research Associations (AiF)
- Bundesanstalt für Landwirtschaft und Ernährung
- Federal Institute for Vocational Education and Training (BIBB)
- The International Bureau (IB) of the BMBF at the DLR
- Deutsches Elektronen-Synchrotron DESY
- German Aerospace Center (DLR)
- EuroNorm Quality Assurance and Innovation Management
- Agency for Renewable Resources (FNR)
- Forschungszentrum Jülich
- Karlsruhe Institute of Technology (KIT)
- Gesellschaft für Anlagen und Reaktorsicherheit (GRS)
- GSI Helmholtz Centre for Heavy Ion Research
- TÜV Akademie Rheinland

[8] Battelle, R&D Magazine, EU R&D Scoreboard.

- VDI Technologiezentrum
- VDI/VDE Innovation und Technik.

The final German supply-side policy for discussion in this text is the direct financing of energy projects. German-financed energy projects played an important role in the rise of solar power in Germany, however, due to the current pace of installation and short-term desire of the German government to slow the pace of solar capacity additions, direct financing of utility-scale solar plants is not expected to be significant over the next several years.

German Demand-Side Policies

Solar Purchase Requirements

Perhaps the most important provision of the EEG is the so-called "connection requirement." The connection requirement mandates preferential treatment for the purchase of grid-connected renewable electricity.

Other Solar Demand-Side Policies

Spain

While Germany has perhaps been the most aggressive nation in climate change policy, Spain's experience is critical in understanding European and global challenges to climate change policy. Its early history of strongly supportive solar policies has given way to a new austerity-based future that has significantly dampened solar optimism.

Spanish Supply-Side Policies

While a smaller overall economy, Spain had perhaps the most aggressive policies to spur solar adoption. Spain set renewable targets in 2008 to achieve 10,000MW of solar power by 2020, launching it to the top of the discussion on favorable solar countries. However, since the financial crisis later that year, Spain moved to cut subsidies for photovoltaics, including the retroactive action of cutting subsidies for projects in progress.[9]

Spain's attempt to meet its renewable energy portfolio was conducted primarily via an aggressive FiT. According to the US Energy Information Agency, "however, without additional controls, generous FiT levels can lead to more investment than intended. One illustration is the Spanish experience, in which the government significantly reduced the tariff a year after its start, and suspended the FiT altogether in 2012, to contain costs to the government and other utility customers."[10]

Since policy certainty and cost containment are two of the most important components necessary for a positive solar environment, the two solar bills that limited—and retroactively removed—solar subsidies were extraordinarily damaging.

[9] Gonzalez, A., Johnson, K., 2009. Spain's solar-power collapse dims subsidy model. Wall Street J.

[10] http://www.eia.gov/todayinenergy/detail.cfm?id=11471.

According to the Spanish solar industry's leading trade body, ASIF, the legislation is "retroactive, discriminatory, and very damaging' to the sector. They will dent the profits of those companies that invested under the previous Spanish regulatory framework…"[11]

As a result, ASIF has filed suit against the Spanish government for violating national and EU regulations, mostly focused on retroactivity.[12] Foreign investors, who lost millions when the market's bottom fell out, have apparently followed suit, hiring attorneys to prepare claims against the Spanish government.[13]

Spain may have concluded that due to its financial crisis, it had no choice but to limit and ultimately remove subsidies and replace them with a 7% energy tax.[14] However, it is clear that the Spanish experience has sent Spain from being a leader in the solar industry to a punchline. According to Phil Dominy of Ernst & Young, who was comparing the FiT reductions in Spain to others in the EU, "Spain stands out as an example of how not to do it."[15]

These Spanish policies have certainly chilled the market and were based on poor planning and overzealous and expensive subsidies. By failing to develop sustainable FiT subsidies, Spain blew into a bubble, racking up a €28 billion ($37.4 billion) tariff deficit. Of interest in other global markets is that the deficit was created by government policy of not only supporting renewables but also keeping electricity prices artificially low. As a result, one energy investor summed it up perfectly when he said "I don't know why anyone would put another penny in investment in the sector in Spain."[16]

Spanish Economic Development Strategies

Recessionary Impacts

As a result of the recession, the Spanish government continues to tighten its belts, looking to reduce spending wherever possible. One way that is severely damaging the industry is by further limiting electric price hikes. "Spanish Prime Minister Mariano Rajoy's government is preparing to cut the regulated revenue of electric utilities and renewable energy generators, threatening their profits as it tries to curb the nation's debt."[17]

This has had a particularly chilling effect on utility-scale solar plants. "Spain's biggest solar power operators are Fotowatio SL and Grupo T-Solar Global SA. Jorge Barredo, president of the solar lobby group UNEF, expects a cut of about €1.2 billion for renewables, including limits to earnings of PV plants. PV plants have been losing

[11] http://www.renewableenergyworld.com/rea/news/article/2011/01/spains-solar-sector-sues-government-over-retroactive-tariff-cuts.

[12] *id.*

[13] http://www.reuters.com/article/2013/02/14/us-spain-renewables-idUSBRE91D1A020130214.

[14] http://www.theglobeandmail.com/report-on-business/international-business/clean-energy-rise-slows-in-europe-as-incentives-yield-to-austerity/article11880892/.

[15] Wilson, P., 2011. Sun setting on European solar subsidies. The Australian.

[16] http://www.reuters.com/article/2013/02/14/us-spain-renewables-idUSBRE91D1A020130214, citing anonymous source.

[17] http://www.businessweek.com/news/2013-06-20/rajoy-targets-endesa-to-edp-s-revenue-to-cut-spain-debt-energy.

€700 million a year in revenue since 2011 because of limits on the hours they can earn subsidies."[18]

Manufacturing and R&D Supportive Policies

The Spanish subsidy programs were initiated to support manufacturers as well as installers; the 44 cent/kWh subsidy applied to manufacturers as well. Spanish Prime Minister Jose Luis Rodriguez Zapatero hoped that the subsidy would create manufacturing jobs in Spain. "Yet by failing to control the program's cost, Zapatero saddled Spain with at least €126 billion of obligations to renewable energy investors. The spending didn't achieve the government's aim of creating green jobs, because Spanish investors imported most of their panels from overseas when domestic manufacturers couldn't meet short-term demand."[19]

Spanish Demand-Side Policies

Solar Purchase Requirements

In March 2006, the Spanish government adopted a new Technical Building Code (CTE, Codigo Tecnico de la Edificacion) which includes a mandate to cover hot water demand with solar thermal energy. The obligation covers all new buildings and those undergoing major renovations. The required solar contribution varies between 30% and 70% depending on three main factors:

- domestic hot water demand of the building (liters/day);
- climate zone;
- conventional fuel to be replaced (only for refurbishments).[20]

While perhaps minor in the overall scheme of solar energy potential in Spain, the thermal water heating policy is at least some positive news for the sector.

Summary

The disappointment for the Spanish solar market is rooted in the depth of optimism that fueled the sector for a decade. Hindsight is always 20–20, but it is hard to fail to recognize that the subsidy programs and artificially depressed electricity prices were a recipe for disaster.

Additionally, Spain's potential as a leading solar marketplace has many industry experts concerned about the future of solar across the globe. "Despite the short-term headwinds that the solar industry faces, the room for growth is almost unlimited. According to data collected by Eurostat [from 2011], Spain's solar energy production as a percent of total electricity production amounted to only 2.13%. This is in comparison

[18] *id.*

[19] http://www.bloomberg.com/news/2010-10-18/spanish-solar-projects-on-brink-of-bankruptcy-as-subsidy-policies-founder.html.

[20] http://www.estif.org/policies/solar_ordinances/; http://www.estif.org/fileadmin/estif/content/policies/STAP/Madrid_SolarRegulation_flyerIDAE_English.pdf.

to Germany's 1.11% and 0.03% in the US. This clearly indicates the large potential in this space."[21] This only compounds the feeling of squandered opportunity in Spain.

Italy

Italy's solar sector has seen "meteoric" growth over the past decade, bolstered by a strong renewable energy portfolio standard and aggressive subsidies. It is estimated that Italy installed as much PV as the United States in 2012, pushing its installed capacity to about 17,000MW by the end of 2012. It also has about 5MW of installed CSP and more in the pipeline.

Such success may lead many to question why Italy dramatically changed its renewable energy policies in 2012. Like its European neighbors, it found the cost of support unsustainable and that consumer exuberance was inflating a potentially dangerous bubble.

Any time actions are taken to mitigate a forming (or existing) bubble, some bursting occurs. Italy saw a future of €9 billion annually for renewable subsidies, costing the average Italian resident over €120 per year. Nearly two-thirds of that cost is to support solar subsidies.[22]

The new program is expected to lower costs for residents and businesses, offering some relief to ratepayers in the EU's highest electricity cost nation. At the same time, it has the solar industry reeling, with investments on hold and experts imploring a wait-and-see, cautionary approach. Pietro Colucci, chairman and chief executive of Italian renewables operator Kinexia, told Reuters that "there is zero interest from foreign investors in greenfield renewable projects [in Italy now]. The new cuts will block decisions to invest since there's just no certainty."[23]

Italy seems to have learned lessons from Spain, Germany, and the rest of the EU, though also misses the mark on several important items. First, it included limits on the amount of annual subsidy. Second, it includes a carbon tax that levels the playing field and internalizes the cost of carbon emissions.

However, one of the key components to successful renewable energy policies is simplicity, and Italy's new scheme is anything but. A combination of fixed, tiered, all-inclusive, and premium FiTs leads to little certainty. Ultimately, as governments move through austerity and resume growth agendas, a premium must be placed on even, fixed, and easy-to-follow subsidy schemes.

Italian Supply-Side Policies

Renewable Portfolio Standard

In the summer of 2012, Italy decided to cut its RPS in favor of FiTs. The new Italian policies commenced at the beginning of 2013. The RPS together with the market-based green certificate program have been scrapped in favor of a new set of policies centered around an FiT scheme and carbon tax. As a result, the Gestore dei Servizi

[21] http://www.dailyfinance.com/2011/05/27/ldk-solar-european-austerity-cloudier-forecast/.

[22] http://www.huffingtonpost.com/2012/04/05/italy-green-energy-incentives_n_1406176.html.

[23] http://www.reuters.com/article/2012/04/05/renewables-italy-idUSL6E8F4C7420120405.

Energetici (GSE) will end its process of buying green certificates by 2015, and utilities will no longer have to purchase certificates under the former RPS policy.[24]

Feed-in Tariff

Italy's RPS has been replaced with a complicated FiT scheme that includes "fixed FiTs, indirect marketing of generation, and a system of premium FiTs available through a tendering process." Altogether, the policies cover a comprehensive list of renewables from solar PV to tidal and wave power. These policies include:

- fixed FiTs (*Conto Energia V*) for solar PV only,
- net metering (*scambio sul posto*) and self-consumption tariffs,
- all-inclusive FiTs (*tariffa omnicomprensiva*) for renewables <1 MW other than solar PV,
- "Premium" model FiTs for all renewables >1 MW other than solar PV, and
- indirect marketing of electricity through GSE (*ritiro dedicato*).[25]

Solar PV is part of a fixed FiT, but it has a hard cap of €6.7 billion and at the time of print, this limit has been nearly achieved. Solar PV systems of <12 kW (typically rooftop residential systems) qualify automatically while larger PV systems (up to 5 MW) must register with GSE to qualify.[26] Projects using net metering, indirect marketing, or state subsidies are excluded from Conto Energia V's all-inclusive tariffs, which are critical because the limits have been nearly met. Tables below indicate the price per kilowatt is significantly <€0.11–0.29 pricing of the FiT, but still more competitive for these alternative mechanisms than the market price of wholesale power.

Conto Energia V First Semester

Photovoltaics	Years	Base Tariff (€/kWh)	1.241 (CAD/kWh)	1.251 (USD/kWh)	Type	Registration
Rooftop all-inclusive						
>1 kW < 3 kW	20	0.208	0.258	0.260	All-inclusive tariff	Direct
>3 kW < 20 kW	20	0.196	0.243	0.245	All-inclusive tariff	Required > 12 kW
>20 kW < 200 kW	20	0.175	0.217	0.219	All-inclusive tariff	Required
>200 kW < 1000 kW	20	0.142	0.176	0.178	All-inclusive tariff	Required
>1000 kW < 5000 kW	20	0.126	0.156	0.158	Premium	Required
>5000 kW	20	0.119	0.148	0.149	Premium	Required
Groundmounted all-inclusive						
>1 kW < 3 kW	20	0.201	0.249	0.252	All-inclusive tariff	Direct
>3 kW < 20 kW	20	0.189	0.235	0.237	All-inclusive tariff	Required > 12 kW

(*continued*)

[24] http://www.renewableenergyworld.com/rea/news/article/2012/12/italy-abandons-rps-adopts-system-of-feed-in-tariffs.

[25] http://www.renewableenergyworld.com/rea/news/article/2012/12/italy-abandons-rps-adopts-system-of-feed-in-tariffs.

[26] http://www.gse.it/en/feedintariff/Photovoltaic/FifthFeed-inScheme/Pages/default.aspx.

Conto Energia V First Semester (Continued)

Photovoltaics	Years	Base Tariff (€/kWh)	1.241 (CAD/kWh)	1.251 (USD/kWh)	Type	Registration
>20 kW < 200 kW	20	0.168	0.208	0.210	All-inclusive tariff	Required
>200 kW < 1000 kW	20	0.135	0.168	0.169	All-inclusive tariff	Required
>1000 kW < 5000 kW	20	0.120	0.149	0.150	Premium	Required
>5000 kW	20	0.113	0.140	0.141	Premium	Required
Innovative integrated rooftop						
>1 kW < 20 kW	20	0.288	0.357	0.360	All-inclusive tariff	Required > 12 kW
>20 kW < 200 kW	20	0.276	0.343	0.345	All-inclusive tariff	Required
>200 kW	20	0.255	0.316	0.319	All-inclusive tariff	Required
Concentrating solar PV						
>1 kW < 20 kW	20	0.259	0.321	0.324	All-inclusive tariff	Required > 12 kW
>20 kW < 200 kW	20	0.238	0.295	0.298	All-inclusive tariff	Required
>200 kW	20	0.205	0.254	0.257	All-inclusive tariff	Required

Source: http://www.gse.it/en/feedintariff/Photovoltaic/FifthFeed-inScheme/Pages/default.aspx.

Italy Ritiro dedicato Indirect Marketing

Limit (kWh)	Min. Tariff (€/kWh)	1.241 (CAD/kWh)	1.251 (USD/kWh)
Wind			
2,000,000	0.076	0.095	0.095
Photovoltaics			
3750	0.100	0.124	0.125
25,000	0.090	0.112	0.113
2,000,000	0.076	0.095	0.095
Geothermal			
2,000,000	0.076	0.095	0.095
Biogas			
2,000,000	0.113	0.140	0.141
Hydro			
250,000	0.150	0.186	0.188
500,000	0.095	0.118	0.119
1,000,000	0.082	0.102	0.103
2,000,000	0.076	0.095	0.095
Biomass			
2,000,000	0.113	0.140	0.141

Generation greater than the limit receives the market price.
Plants using this model cannot use fixed premium or net metering.
Source: http://176.9.160.135/search-by-country/italy/single/s/res-e/t/promotion/aid/feed-in-tariff-ii-ritiro-dedicato/lastp/151/.

Conto Energia V First Semester 2013 Self-Consumption Premium

Photovoltaics	Years	Base Tariff (€/kWh)	1.241 (CAD/ kWh)	1.251 (USD/ kWh)	Type	Registration
Rooftop all-inclusive						
>1 kW < 3 kW	20	0.126	0.156	0.158	All-inclusive tariff	Direct
>3 kW < 20 kW	20	0.114	0.141	0.143	All-inclusive tariff	Required > 12 kW
>20 kW < 200 kW	20	0.093	0.115	0.116	All-inclusive tariff	Required
>200 kW < 1000 kW	20	0.060	0.074	0.075	All-inclusive tariff	Required
>1000 kW < 5000 kW	20	0.044	0.055	0.055	Premium	Required
>5000 kW	20	0.037	0.046	0.046	Premium	Required
Groundmounted all-inclusive						
>1 kW < 3 kW	20	0.119	0.148	0.149	All-inclusive tariff	Direct
>3 kW < 20 kW	20	0.107	0.133	0.134	All-inclusive tariff	Required > 12 kW
>20 kW < 200 kW	20	0.086	0.107	0.108	All-inclusive tariff	Required
>200 kW < 1000 kW	20	0.053	0.066	0.066	All-inclusive tariff	Required
>1000 kW < 5000 kW	20	0.038	0.047	0.048	Premium	Required
>5000 kW	20	0.031	0.038	0.039	Premium	Required
Innovative integrated rooftop						
>1 kW < 20 kW	20	0.186	0.231	0.233	All-inclusive tariff	Required > 12 kW
>20 kW < 200 kW	20	0.174	0.216	0.218	All-inclusive tariff	Required
>200 kW	20	0.153	0.190	0.191	All-inclusive tariff	Required
Concentrating solar PV						
>1 kW < 20 kW	20	0.157	0.195	0.196	All-inclusive tariff	Required > 12 kW
>20 kW < 200 kW	20	0.136	0.169	0.170	All-inclusive tariff	Required
>200 kW	20	0.103	0.128	0.129	All-inclusive tariff	Required

Source: http://www.gse.it/en/feedintariff/Photovoltaic/FifthFeed-inScheme/Pages/default.aspx.

Net Metering

Net metering is an important alternative to the FiT. Net metering essentially offers the owner of the system "credits" to use towards their energy bills if excess energy they produce is fed back into the grid, essentially rolling back the meter. This simple mechanism allows a reduction essentially at retail rates, however, does not incentivize systems larger than a site's own needs.

According to Italian industry experts, "the revised scheme has simplified procedures for the calculation of kilowatt hour credits and will be limited to PV systems no greater than 200 kW. Without the feed-in tariff to provide that extra push, the net metering system could be set to play a more important role in the Italian solar market in driving solar uptake as well as self-consumption."[27]

In addition to the limited subsidies, Italy is considering a carbon tax with proceeds earmarked for financing renewable energy production in a dramatic shift in clean energy financing. As the measure sits in front of parliament, Italy is also pushing the EU to focus on a carbon tax for both emissions and products that lead to increased emissions. The government is also considering imposing excise duties on energy products based on carbon content (similar to an EU scheme, and "will coordinate the launch of the environmental fiscal measures with other European countries,' the government said in a statement."[28]

Italian Economic Development Strategies

Recessionary Impacts

Italian solar firms have felt the same pain as other companies during the economic crisis. In a nutshell, capital expenditures have dropped due to falling component prices, borrowing costs are up due to higher interest rates, and the market has declined due to policy uncertainty and lower energy prices fueled by lower economic output. As the business development manager of Unicredit Leasing put it, the economic crisis has had an impact on borrowing costs: "Recently we faced a decrease in base reference interest rates but also a higher rise of liquidity costs which drove a significant increase on overall borrowing costs." But he adds that within the renewable energy sector, this increased cost is counterbalanced by a decrease in capital expenditure costs—especially for PV, but also for wind and bioenergy. "Making an investment in renewable here seems to be still profitable and is welcomed from our side."[29]

Growth Versus Austerity

As indicated by Italy's pronouncement that it will no longer subsidize solar after the €6.7 billion cap is hit, the austerity budget hawks have won the battle. With <€100 million left in subsidies remaining, the Italian solar industry is basically on its own.[30] Ultimately, time will tell whether these austerity measures end up helping the sector to become more efficient and competitive, or doom it to failure.

Ultimately, Italy has attempted to slowly deflate its emerging solar bubble and may have found a successful, if complicated model for the future. Attractive financing and continued declines in module prices will be critical for the industry to continue its growth. As noted by Paolo Gianese, general secretary of the Italian solar

[27] http://www.pvtech.org/friday_focus/friday_focus_the_future_of_the_unsubsidised_italian_solar_market.
[28] Reporting by Svetlana Kovalyova, editing by David Cowell. *Source*: http://www.reuters.com/article/2012/04/17/italy-carbontax-idAFL6E8FHALR20120417.
[29] http://www.renewableenergyworld.com/rea/news/article/2013/06/can-italy-keep-its-renewables-investors.
[30] http://www.pv-tech.org/friday_focus/friday_focus_the_future_of_the_unsubsidised_italian_solar_market.

association Industrie Fotovoltaiche Italiane (IFI), "the sector needs sustainability and predictability for a safe growth of the sector. Stop and go measures create only market disruptions. This means that it is better to set low incentives first, in order not to create high speculation as we experienced in Italy. We don't know who will be next, may be some country from Eastern Europe."[31]

China

China is one of the most important emerging solar markets, and as a (at least substantially) managed economy, policy is an even more important driver there than in other nations. Its predominant policies have been to drive production, but with the bottom of the module market falling out and an apparent lack of appetite for continued production stimulus, implementation policies will only grow in importance.

Chinese Supply-Side Policies

Renewable Portfolio Standard

China adopted a renewable energy target in 2006 and modified it in 2009 to the following targets:

- Renewable electricity: 500 GW by 2020 (300 from hydro, 150 from wind, 30 from biomass, and 20 from solar PV).
- Renewable energy: 15% by 2020.[32]

China has adopted renewables at a rapid pace, though its history-making economic growth has led to skyrocketing energy usage. Therefore, the energy mix has not changed as dramatically as might be expected from the addition of new renewable systems.

Total clean energy production was up from 7.5% to 9% between 2005 and 2008, and the solar PV capacity added each year has risen.

Photovoltaics

Year	Capacity (MW)	Installed
2000	3	
2001	4.5	
2002	18.5	
2003	10	
2004	10	
2005	8	
2006	10	

(*continued*)

[31] http://www.pv-tech.org/friday_focus/friday_focus_the_future_of_the_unsubsidised_italian_solar_market.
[32] Martinot, E., Junfeng, L., 2010. Renewable energy policy update for China. Renewable EnergyWorld (retrieved 14.11.10).

Photovoltaics (Continued)

Year	Capacity (MW)	Installed
2007	20	
2008	40	
2009	160	
2010	500	
2011	2500	
2012	5000	

Source: National Survey Report of PV Power Applications in China 2011.

Feed-in Tariff

China adopted a new FiT policy named Notice on Perfection of Policy Regarding Feed-in Tariff of Power Generated by Solar PV (国家发改委关于完善太阳能光伏发电上网电价政策的通知) in 2011. In that new policy, PV is broken into two groups. The first, prior to Q3 2011 and in operation by the end of 2011, received a tariff of about $0.18 (RMB 1.15). Those following July 1, 2011 have a slightly lower FiT of RMB 1.0 ($0.15) per kilowatt hour.[33]

The new policy gives the energy regulator sweeping power to adjust future rates in order to address changing market conditions. This type of flexibility is rare in other countries due to the political challenges facing any changes to renewable energy policy. With this flexibility, China will be able to respond much more quickly to issues with its account balance, technology, and cost of capital. Finally, the new FiT policy mandates that projects that are put to bid cannot exceed the maximum FiT level. [34]

Despite these changes, China seems to have learned from Europe (and perhaps its own past) by continuing to make payments under its previous solar program, called building integrated PV (BIPV). This previous policy, enacted in 2009, offered varied subsidies, and the new FiT adds a lower premium to reflect that. Specifically, FiT payments are connected to desulfurized coal projects, which range from RMB 0.25 to RMB 0.50.[35]

According to Chinese industry experts, the new FiT policy does not address several fundamental solar issues in the country. Namely, these include:

- "The existing rules fail to mention a time period for the FiT. This may be intentional by NDRC—to give it room for adjustment. As discussed above, NDRC has the right to adjust tariff rates depending on the availability of investment capital and technological advances, which we believe suggests that if the policy fails to attract the desired number of participants, NDRC may consider increasing the tariff next year to make it more attractive. Notably, NDRC took the same strategy in its FiT plans for wind power.
- Only one FiT rate is offered for all solar PV projects, regardless of the region or installation method. Developers may rush to provinces that have a rich solar resource, such as the

[33] However, exceptions have been given to projects located in Tibet, which, under certain circumstances can still receive a FiT of RMB1.15. http://www.mondaq.com/x/159390/Renewables/China+Policy+Shedding+Light+On+The+Recently+Enacted+Solar+FeedInTariff.

[34] *id.*

[35] *id.*

provinces of Qinghai, Tibet, and Xinjiang. Since these relatively remote areas are located far away from demand, their connection to major energy grids represents an issue of significant concern. Improved coordination will be required between the solar PV industry, provincial governments, and electric grid corporations. Ironically, the same issue was identified and, to some extent, solved in the wind FiT policy promulgated by NDRC in 2009. NDRC divides China's territory into four sections based on the level of wind force, and each part enjoys a different tariff rate. If NDRC does not implement a similar solution through future solar power regulations, developers and grid companies should prepare carefully to overcome the issues raised from a uniform solar tariff.

- Grid connection and transmission issues remain key concerns. Along with the expected rapid expansion of solar power generation capacity, the transmission capacity of the national grid must grow to meet the anticipated growth of China's solar PV power stations. The same issue was, and continues to be, a vexing issue for China's wind industry and remains unsolved: only about 70% of China's total wind power capacity is connected to the grid. As mentioned above, much of the solar energy from PV facilities may need to be sent thousands of miles away to the power-hungry provinces in eastern China. The solar market will likely face similar if not identical grid connection and transmission constraints as wind power. Regrettably, neither the current solar FiT policy nor any other renewable energy laws and regulations address the issue."[36]

Another important program that appears due for revision is the Golden Sun program. This program has been viewed as a failure to produce significant amounts of power. At the same time, rumors persist that the entire FiT program will be scrapped and new more efficient policies based on distributed generation will replace them.[37] As of June 2013, some optimists expect Beijing to increase higher subsidies and create new ones for household power.[38]

Consumer Incentives and Rebates

China has an incredible ability to influence its market. This pricing control, if applied to domestic photovoltaics, could be game changing for the industry. Recently, a report commissioned by the National Renewable Energy Laboratory (NREL) demonstrates this ability with water heating. In the United States, solar water heaters can cost up to $10,000 to install, while in China, it can be as cheap as $300.[39]

Chinese Economic Development Strategies

Recessionary Impacts

The global recession shifted the solar industry's trajectory in many ways. One of the most dramatic is that it resulted in a tilting of renewable energy investment (particularly in the public sector) away from developed countries and towards developing

[36] http://www.mondaq.com/x/159390/Renewables/China+Policy+Shedding+Light+On+The+Recently+Enacted+Solar+FeedInTariff.

[37] http://www.gsb.stanford.edu/news/headlines/chinas-solar-panel-boom-bust.

[38] http://english.people.com.cn/90778/8273874.html.

[39] http://www.eenews.net/stories/1059983772.

ones. Of the $143 billion in global renewable investments in 2010, China attracted $48.5 billion or more than one-third of the total global investment. The developing world is clearly catching up in terms of its policies, capital invested, manufacturing, and maturation of markets.[40]

At the same time, China is responsible—good or bad—for the oversupply of solar modules. The oversupply has been driven by overinvestment and market manipulation that has so drastically cut prices and profit margins. As noted previously, the decline in prices is the bane of global (and even now, Chinese) manufacturers but the boon for the installation sector. Bankruptcies have already impacted the landscape in China and across the rest of the globe.[41]

2012 was a bad year for Chinese solar firms. After continuing an impressive growth rate to approximately 900 firms by the end of 2011, nearly all manufacturers, the number of firms in business dropped dramatically in 2012 to approximately 700. This was the result of about 100 new firms being created and 300 going out of business. In China, where the government is quickly losing its appetite to prop up failing solar manufacturers, the pace of consolidation and closure continues to quicken.[42,43]

The resulting landscape also has major players in China, such as LDK, Suntech, and Yingli, either in bankruptcy or nearing it. These firms grew rapidly during the solar boom, and used debt financing to fuel their growth. The resulting price collapse and reduced profit margins make it impossible for the firms to continue to manage their extreme debt loads.

Unlike in the United States or EU, however, where downsizing and consolidation are commonplace responses to market collapse, local governments have been heavily involved in keeping the companies afloat in order to keep their workers employed. This is despite diminished support from Beijing. With prices continuing to fall, however, it is unclear that this strategy is working, and downsizing seems inevitable.[44]

Chinese Demand-Side Policies

China seems to recognize that it must develop a domestic market in order to relieve the market of its oversupply. With the policy uncertainty and continued economic malaise embroiling Europe, China must boost its own solar demand in order to save its manufacturing core. The government response will likely be heavily geared towards the domestic PV market, which is expected to increase to 10GW in 2013. The following incentives are currently being explored:

1. Encourage the domestic distributed solar PV project deployment.
2. State grid companies are urged to buy all available solar power, and power transmission facilities must be built to accommodate solar power generation.

[40]http://www.worldwatch.org/renewable-energy-continued-growth-2010-despite-recession.

[41]http://www.greentechmedia.com/articles/read/Rest-in-Peace-The-List-of-Deceased-Solar-Companies.

[42]ENF Market Survey, Chinese Cell and Panel Manufacturers, 2012.

[43]These woes come on the heels of the filing of trade sanctions by the United States and EU against China for its alleged market manipulation. This, in turn, has led China to file sanctions against the United States for its practices in selling polysilicon (among other, nonsolar items). The chilling effect of these cases is yet to be known.

[44]http://thediplomat.com/pacific-money/2012/12/08/lights-out-for-chinas-solar-power-industry/.

3. Regulate solar power pricing and expand national renewable energy funding to subsidize solar power generation in time.
4. Financial institutions are convinced to support PV companies and help them overcome financing difficulties.
5. Strengthen research and development into key materials and equipment in the PV industry.
6. Control excess production capacity expansion and bring up the competitive companies through mergers and reorganizations among PV makers.[45]

The Chinese government is therefore focusing on PV installation and downstream pulling rather than its past policy of supporting production capacity expansion yet despite the promise of deployment, significant obstacles remain.

2013 was to be a banner year for Chinese domestic PV installations with new policies to encourage the use of smaller scale, rooftop mounted systems for home use. Thousands of interested Chinese flocked to the State Grid Corporation to learn more about solar on their rooftops. However, as of mid-May 2013, only 25 home PV power stations are in the process of installation across Jiangsu province with a total installed capacity of 154.7 kW.[46]

The obstacles to installation are significant. First, prospective owners must get their neighbor's permission before panels can be placed on the roof. [47] This appears to be a challenge not dissimilar to the experience of US residents living in communities with homeowners association regulations (such as condominiums and gated communities).

Even once the permission is granted, there is virtually no market of installers yet in China. As a result, buyers are expected to locate, purchase, and install the systems themselves, which is complicated and dangerous. While there are a few firms that sell and install systems in China for the domestic market, the immaturity of that market has created a clear chicken-and-egg scenario, where installers do not want to enter into a market with no consumers and consumers cannot enter a market without installers.[48]

Perhaps the greatest difficulty facing the widespread adoption of household PV is the low price of electricity, which lengthens the ROI horizon. As reported in Chapter 9, one of the key components to solar cost competitiveness is the utility retail price of electricity. At 0.5 yuan or $0.08/kWh, the ROI, even with very low capital costs of about $7500 for a 3 kW system, approaches 20 years.[49]

Ultimately, the policies to spur domestic rooftop solar installation are too few and too weak. While they may be an appropriate mechanism for market forcing and are undoubtedly developed with an eye to success and failure in the West, China is in need of a much bolder policy to *create* a solar market. Strong incentives, higher, longer term price supports, and more supportive policies are necessary to convince Chinese companies to jump into the domestic market and reduce the payback time to spur Chinese consumers.

[45] http://seekingalpha.com/article/1533742-chinese-solar-photovoltaic-manufacturers-no-worse-for-wear.
[46] http://www.renewableenergyworld.com/rea/news/article/2013/07/residential-solar-pv-systems-experiencing-slow-adoption-in-china.
[47] *id.*
[48] *id.*
[49] *id.*

The Chinese government seems to have gotten the message and is working towards some solutions, even if they are still too small to be market creating. One such innovative idea is a plug-and-play device that can be installed on a balcony or window. Much more importantly, Beijing's National Development and Reform Commission appears to be honing in on a subsidy scheme for distributed PV that would make the systems more attractive. This will be a critical component to creating the industry, and if successfully implemented, clearing the glut of panels, further spurring their manufacturing core.

6 Federal and State Energy Policies

The United States is far behind its competitors in national-level solar policy, which is expected given that it has never had a comprehensive energy policy. While the federal consumer tax credit of 30% for solar systems is one of the strongest drivers of the industry, the incentive structure from the US government is more of a patchwork quilt than a comprehensively and cohesively developed plan. Unlike its rivals, it has no national feed-in-tariff (FIT) legislation and no renewable portfolio standard.

This chapter is a summary of policies at the state and federal level which are subject to consistent change. While the information represents the author's best attempt to organize existing information, solar professionals should not rely on the information for decision-making, and should consult tax and legal professionals regarding the application of specific provisions.

Federal Policies

The United States does have a number of important policies and incentives, however, though many are set to expire in the near term and most function to reduce costs rather than to support markets for the energy (see the contrast with Germany in Chapter 5).

Tax Credits

Personal Tax Credit

Perhaps the most important federal solar policy in the United States—and certainly the most important to the residential market—is the personal tax credit for renewable energy systems. This credit, initially enacted in 2005, allows taxpayers to take a credit of up to 30% of the installed cost of a solar-electric or hot water system. The bill, which initially included a cap of $2000 and was not applicable to Alternative Minimum Tax (AMT) payers, was amended and extended as part of the American Recovery and Reinvestment Act (ARRA, or "the stimulus bill").

The ARRA revision removed the $2000 cap, allowed AMT payers to claim the credit, and extended its provisions through 2016. This last piece has been important to provide stability, however, if it is not reenacted, expect 2016 to be a bumper year for residential solar (to maximize the credit) followed by a substantial downturn in 2017. Despite price declines, the 30% tax credit is an integral piece to the affordability of solar systems. Its extension is crucial to the continued success of the residential market.

Solar Energy Markets. DOI: http://dx.doi.org/10.1016/B978-0-12-397174-6.00006-4

Investment Tax Credit

Similar to the personal tax credit, the corporate Business Energy Investment Tax Credit (ITC) allows for a 30% credit with no limit for most solar technologies. Approved applications include solar-electric systems and water heating, though pool heating and passive solar systems are excluded (similar to the personal credit). It is similarly set to expire on December 31, 2016, after being extended and amended several times since its enactment; however, the ITC will not disappear but the credit for solar-electric systems will drop from 30% to 10%.[1]

Advanced Energy Manufacturing ITC

Rounding out the tax credits is the Advanced Energy Manufacturing ITC. This incentive allows a tax credit of up to 30% of qualified investments in creating, retooling, or expanding manufacturing facilities that produce clean energy products. The $2.3 billion program has a cap of $30,000, with $150 million allotted for 2013.

The credit applies to tangible property only (excluding buildings and other structures) and is typically used for investments in equipment. To receive funds, manufacturers apply for the credit. The US Department of Energy makes recommendations to the US Treasury Department, which issues the certifications. Recipients have 1 year to meet all requirements and must have the equipment in service within 3 years from issuance. Importantly, business that claim this credit may not also take the ITC.[2]

Modified Accelerated Cost Recovery System

One of the key federal incentives—particularly for larger scale systems and third-party financed leases—is the Modified Accelerated Cost Recovery System (MACRS). This provision allows corporations to accelerate the depreciation of property (in this case, solar systems) over a 5-year period. This depreciation can be a significant difference maker in third-party leases because companies receive significant depreciation of the property over the short term and pass the savings on to consumers.

While this system has been in place since 1986, recent years have had a series of bonus appreciation amendments. These amendments, beginning in 2008 and running through 2013 (so far), allow companies to deduct 50% of the property's depreciation in the first year. This provision significantly reduces the up-front cost to consumers and allows for more rapid payback of solar systems.[3]

North Carolina State University's Database of State Incentives for Renewable Energy (DSIRE) program notes this important caution: "The bonus depreciation rules do not override the depreciation limit applicable to projects qualifying for the federal business energy tax credit. Before calculating depreciation for such a project, including any bonus depreciation, the adjusted basis of the project must be reduced by one-half of the amount of the energy credit for which the project qualifies."[4]

[1] http://www.dsireusa.org/incentives/incentive.cfm?Incentive_Code=US02F&re=1&ee=1.

[2] http://www.dsireusa.org/incentives/incentive.cfm?Incentive_Code=US52F&re=1&ee=1.

[3] *id.*

[4] *id.*

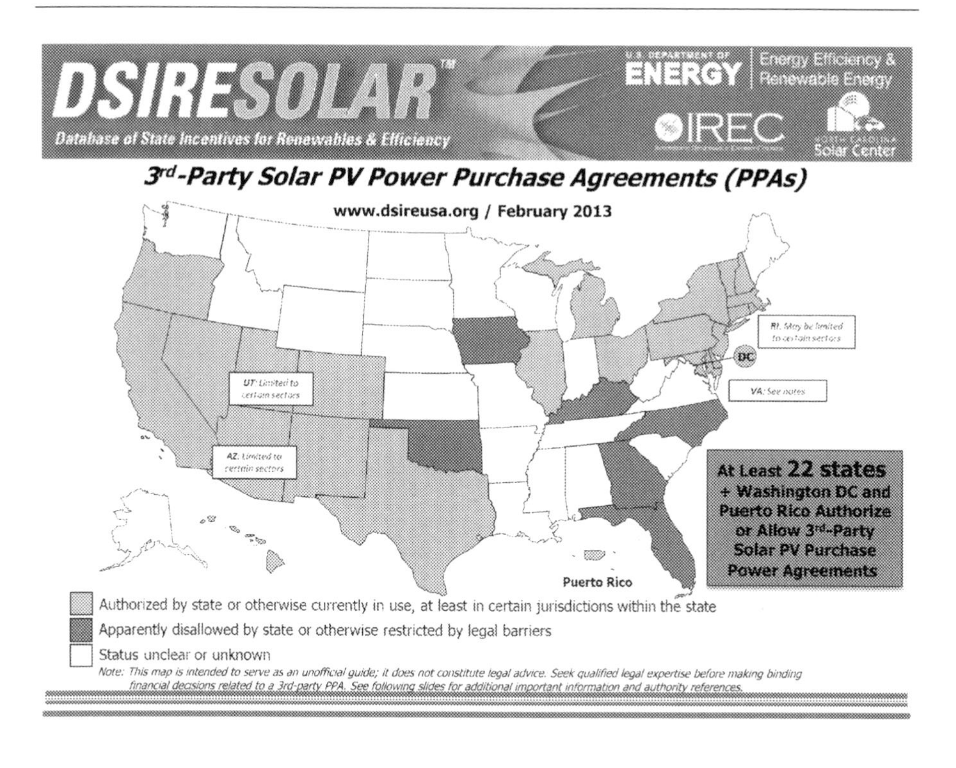

Important Information Regarding 3rd-Party Solar PPAs

Authorization for 3rd-party solar PV PPAs usually lies in the definition of a "utility" in state statutes, regulations or case law; in state regulatory commission decisions or orders; and/or in rules and guidelines for state incentive programs.

Even though a state may have authorized the use of 3rd-party PPAs, it does not mean that these arrangements are allowed in every jurisdiction. For example, municipal utilities may not allow 3rd-party PPAs in their territories even though they are allowed or in use in the state's investor-owned utility (IOU) territories.

Though a 3rd-party PPA provider may not be subject to the same regulations as utilities, additional licensing requirements may still apply.

This map and information is provided as a public service and does not constitute legal advice. Seek qualified legal expertise before making binding financial decisions related to a 3rd-party PPA.

DSIRE acknowledges IREC and Keyes, Fox & Wiedman LLP, for their support in creating and maintaining this resource.

This information is updated quarterly or as new information is verified.

Please send comments or questions to Chelsea Barnes at chelsea_barnes@ncsu.edu.

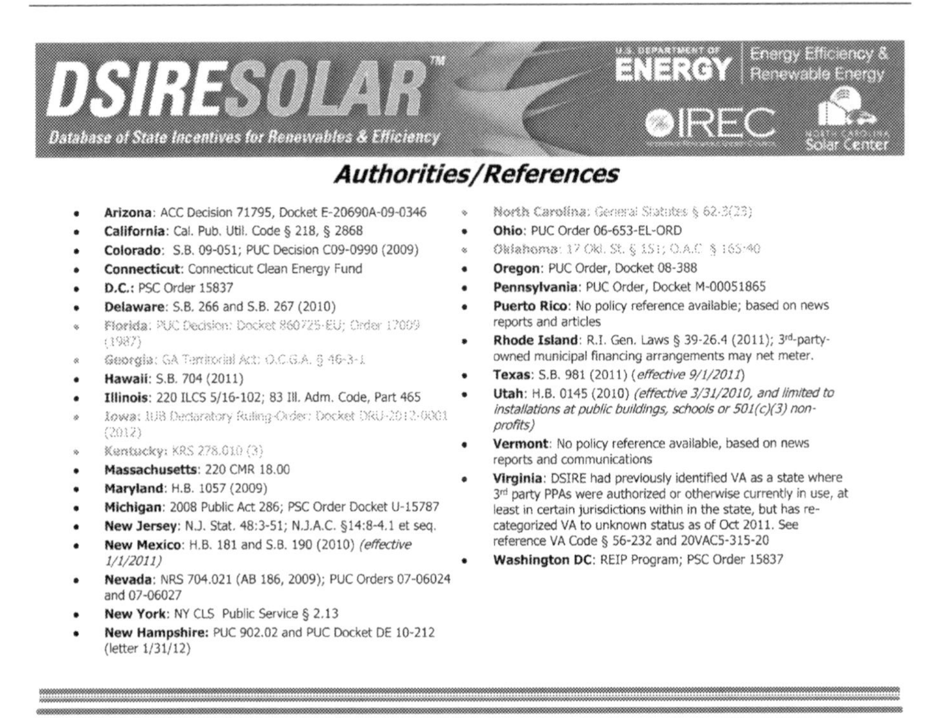

Authorities/References

- **Arizona**: ACC Decision 71795, Docket E-20690A-09-0346
- **California**: Cal. Pub. Util. Code § 218, § 2868
- **Colorado**: S.B. 09-051; PUC Decision C09-0990 (2009)
- **Connecticut**: Connecticut Clean Energy Fund
- **D.C.**: PSC Order 15837
- **Delaware**: S.B. 266 and S.B. 267 (2010)
- **Florida**: PUC Decision: Docket 860725-EU; Order 17009 (1987)
- **Georgia**: GA Territorial Act: O.C.G.A. § 46-3-1
- **Hawaii**: S.B. 704 (2011)
- **Illinois**: 220 ILCS 5/16-102; 83 Ill. Adm. Code, Part 465
- **Iowa**: IUB Declaratory Ruling-Order: Docket DRU-2012-0001 (2012)
- **Kentucky**: KRS 278.010 (3)
- **Massachusetts**: 220 CMR 18.00
- **Maryland**: H.B. 1057 (2009)
- **Michigan**: 2008 Public Act 286; PSC Order Docket U-15787
- **New Jersey**: N.J. Stat. 48:3-51; N.J.A.C. §14:8-4.1 et seq.
- **New Mexico**: H.B. 181 and S.B. 190 (2010) *(effective 1/1/2011)*
- **Nevada**: NRS 704.021 (AB 186, 2009); PUC Orders 07-06024 and 07-06027
- **New York**: NY CLS Public Service § 2.13
- **New Hampshire:** PUC 902.02 and PUC Docket DE 10-212 (letter 1/31/12)
- **North Carolina**: General Statutes § 62-3(23)
- **Ohio**: PUC Order 06-653-EL-ORD
- **Oklahoma**: 17 Okl. St. § 151; O.A.C. § 165:40
- **Oregon**: PUC Order, Docket 08-388
- **Pennsylvania**: PUC Order, Docket M-00051865
- **Puerto Rico**: No policy reference available; based on news reports and articles
- **Rhode Island**: R.I. Gen. Laws § 39-26.4 (2011); 3rd-party-owned municipal financing arrangements may net meter.
- **Texas**: S.B. 981 (2011) *(effective 9/1/2011)*
- **Utah**: H.B. 0145 (2010) *(effective 3/31/2010, and limited to installations at public buildings, schools or 501(c)(3) non-profits)*
- **Vermont**: No policy reference available, based on news reports and communications
- **Virginia**: DSIRE had previously identified VA as a state where 3rd party PPAs were authorized or otherwise currently in use, at least in certain jurisdictions within in the state, but has re-categorized VA to unknown status as of Oct 2011. See reference VA Code § 56-232 and 20VAC5-315-20
- **Washington DC**: REIP Program; PSC Order 15837

Grants

While the most prolific and widely applicable US federal solar incentives are tax credits and deductions, there are several targeted grant programs that offer direct cash payments to solar energy producers.

Rural Energy for America Program Grants

The United States Department of Agriculture (USDA) administers a direct grant program called the Rural Energy for America Program (REAP). The grants, ranging from $2500 to $500,000, can cover up to 25% of total project costs, and are designed to help agricultural producers and rural small businesses to reduce the cost of their energy by defraying system costs. Of specific importance to this grant, land grant and other universities may be eligible for funding (which typically cannot participate in tax credit schemes due to their nonprofit status), though K-12 schools are not eligible.[5]

USDA also has offered a High Energy Cost Grant Program since 2000. According to DSIRE, "The U.S. Department of Agriculture (USDA) offers an ongoing grant program for the improvement of energy generation, transmission, and distribution facilities in rural communities. This program began in 2000. Eligibility is limited to projects in communities that have average home energy costs at least 275% above the national average. Individuals, non-profits, commercial entities, state and local governments

[5] http://www.dsireusa.org/incentives/incentive.cfm?Incentive_Code=US05F&re=1&ee=1.

(including any agency or instrumentality thereof), and tribal governments are eligible for this grant. Individuals must work on a project that will benefit the community in order to qualify. Under the most recent solicitation for projects, a total of $7 million was available for qualifying projects."[6] While the most recent grant application closed on 2012, with a grant range of $20,000 to $3 million, USDA is likely to continue the program in the future. A loan guarantee fund is also part of the program.

Tribal Energy Program Grant

The United States Department of Energy offers several different grants to subsidize the cost of solar heating and cooling and solar-electric production on federal tribal lands. These measures vary frequently over time and solicitations can be tracked at: http://apps1.eere.energy.gov/tribalenergy/financial_opportunities.cfm.

Loans and Loan Guarantees

Another popular federal solar incentive program involved direct loans and loan guarantees. They are also controversial, after Solyndra filed for bankruptcy and failed to repay its loan.

Department of Energy Loan Guarantee Program

The Department of Energy developed a loan guarantee program in 2005, called the 1703 program. As part of ARRA, this program was expanded by adding a section 1705. The 1705 program expired in 2011, but the original authority to provide up to $10 billion in energy-related projects under 1703 remains.

The program specifically focuses on manufacturing facilities, renewable energy projects, and integrated projects. Repayment terms are quite favorable, with a maximum period of 30 years or 90% of the assets' expected useful life.[7]

Clean Renewable Energy Bonds

Clean Renewable Energy Bonds (CREBs) were a popular funding mechanism that were similar to other forms of bond financing, though the credits are subject to income tax unlike many public bonds. The federal CREB program is no longer accepting applications (administered by the Department of the Treasury) but many states do.

Qualified Energy Conservation Bonds

According to the Center for Sustainable Energy, "Qualified Energy Conservation Bonds (QECBs) can be used by local and state government agencies on a wide range of activities; nationwide the QECB program is funded at $2.4B; each state has received an allocation based on population, and each state must develop its own mechanism for how allocations are awarded to individual projects. The ARRA authorized local communities

[6] http://www.dsireusa.org/incentives/incentive.cfm?Incentive_Code=US56F&re=1&ee=1.

[7] http://www.dsireusa.org/incentives/incentive.cfm?Incentive_Code=US48F&re=1&ee=1.

to use some or all of their QECB allotment for funding municipal solar and energy efficiency projects, including capital expenditures that reduce energy consumption on publicly-owned buildings by at least 20%, and implementing green community programs. Such 'green community programs' could include many opportunities for cities, which could issue QECBs to fund loan, rebate and/or grant programs…."[8]

QCEBs function by providing a tax credit to the bondholder in lieu of the issuer paying interest to the bondholder. This is therefore equivalent of interest-free financing. Of current and specific use for solar activities, many municipalities and states are issuing QCEBs for community-based solar.

QCEBs are different from CREBs because they are not subject to the Department of Treasury approval process. Rather, they are based on each state's population as of July 1, 2001, and then states allocate to local governments based on their size as well.[9]

State Policies

The lack of a national renewable energy policy has led many states to act on their own. According to The Solar Foundation's Solar Jobs Census, state legislation, including renewable portfolio standards and third-party ownership allowance, was the second most important factor for their growth.

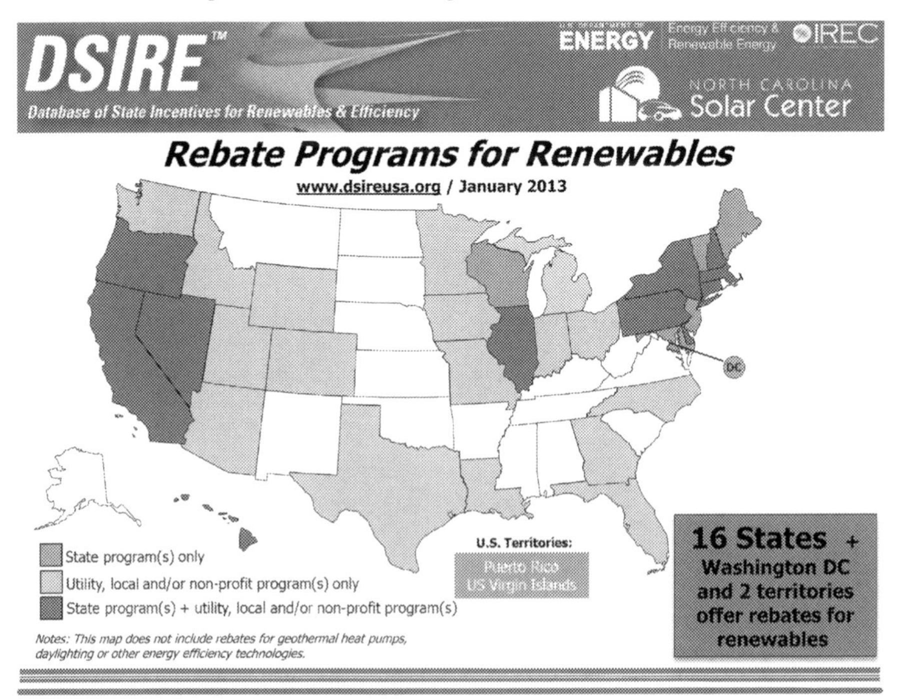

[8] http://energycenter.org/index.php/public-affairs/federal-legislation/1283-qualified-energy-conservation-bonds-qecbs.

[9] http://www.dsireusa.org/incentives/incentive.cfm?Incentive_Code=US51F&re=1&ee=1.

This section identifies and explains various state-level incentives, while providing examples of activity from around the country. The DSIRE[10] is the most comprehensive source of up-to-date state-level incentives, available on the Internet at http://dsireusa.org.

Renewable Portfolio Standards

As cited previously in this text, a renewable portfolio standard is an important forcing tool used by states to spur the deployment of clean energy, and most frequently, solar power. The United States is far behind its competitors with its lack of a national standard, but many states have filled the gap. The below map illustrates the states with an adoptedrenewable portfolio standard, and is followed by another map, which shows those with a specific solar requirement.

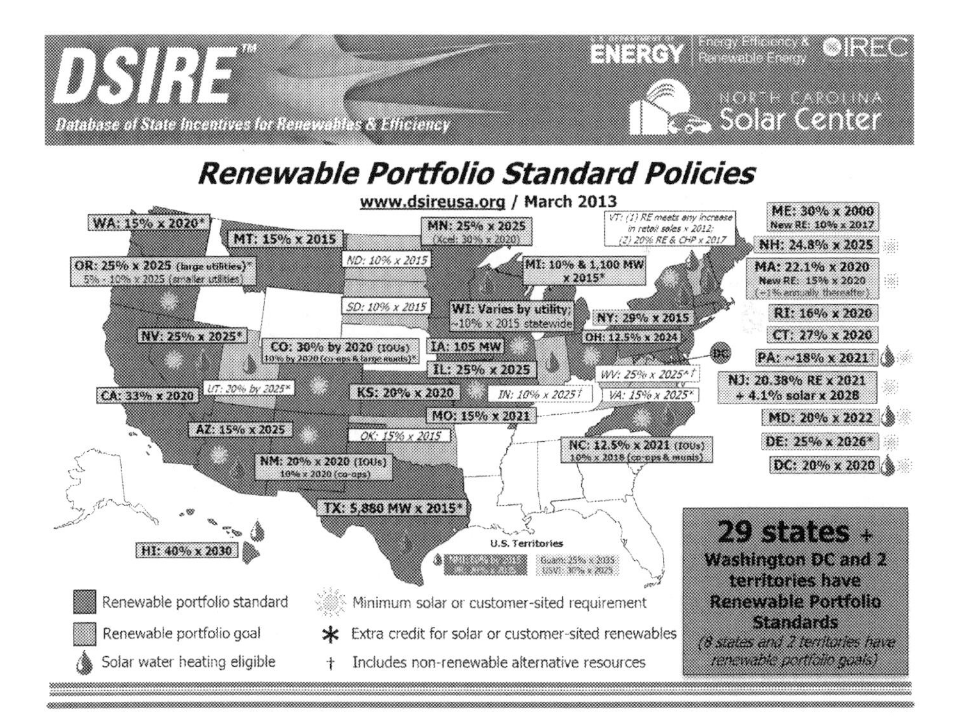

[10]Established in 1995, DSIRE is currently operated and funded by the NC Solar Center at NC State University, with support from the Interstate Renewable Energy Council, Inc. DSIRE is funded in part by the US Department of Energy. DSIRE data provides the basis for this section.

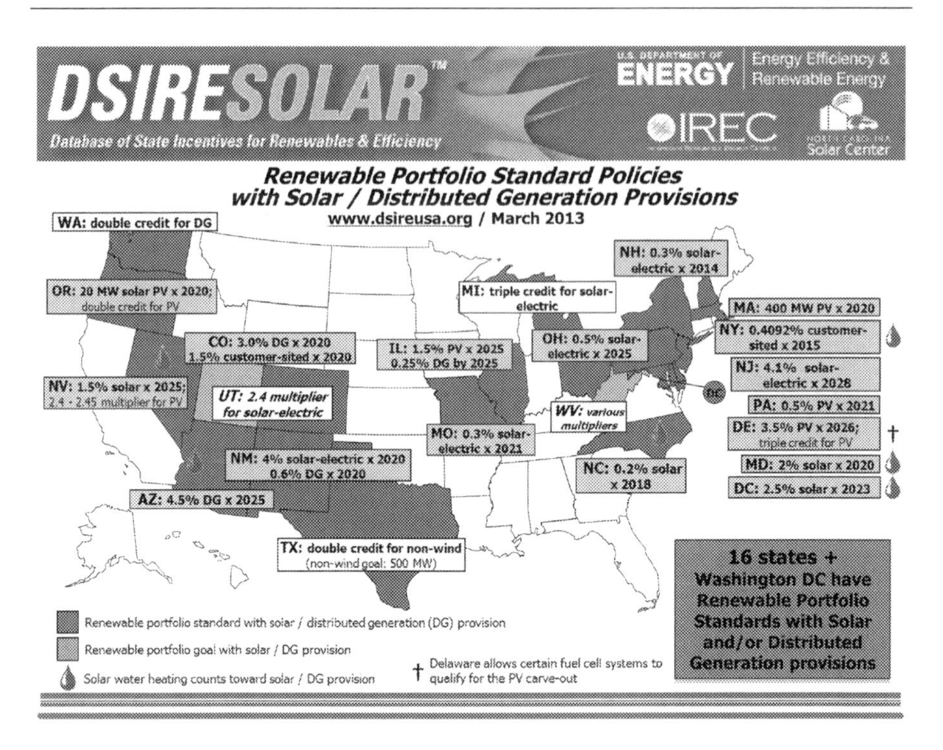

Tax Credits

Tax credits are the most popular US solar incentive. A tax credit simply refers to a reduction in tax liability, typically at the state or local level (though some municipalities and counties that tax income also offer credits). The primary benefit of a tax credit for consumers are reduced overall pricing (and though delayed until tax filing) and lower up-front costs. While the state and municipal tax credits are generally not enough to have dramatic impacts on the cost of the systems, it is often enough to push potential buyers who are on the fence.

States often prefer tax credits because they reduce receipts rather than increase expenditures. While this has the same end result on the treasury, they differ in at least three important ways. First, tax credits do not require the same administration. System costs can often be simply included on a tax return. Second, funding does not need to be allocated. It is a hit on revenue, not a new expense. Third, it is more politically palatable. It is much easier for a politician to vote in a tax cut as opposed to approving new spending. Especially in this time of state and federal budget austerity, many states do not have the ability to directly fund cash or other incentives, so a tax credit is a more viable option to promote the industry.[11]

[11]Gouchoe, S., Everette, V., Haynes, R., 2002. Case Studies on the Effectiveness of State Financial Incentives for Renewable Energy. North Carolina Solar Center and National Renewable Energy Laboratory (NREL), Raleigh, NC.

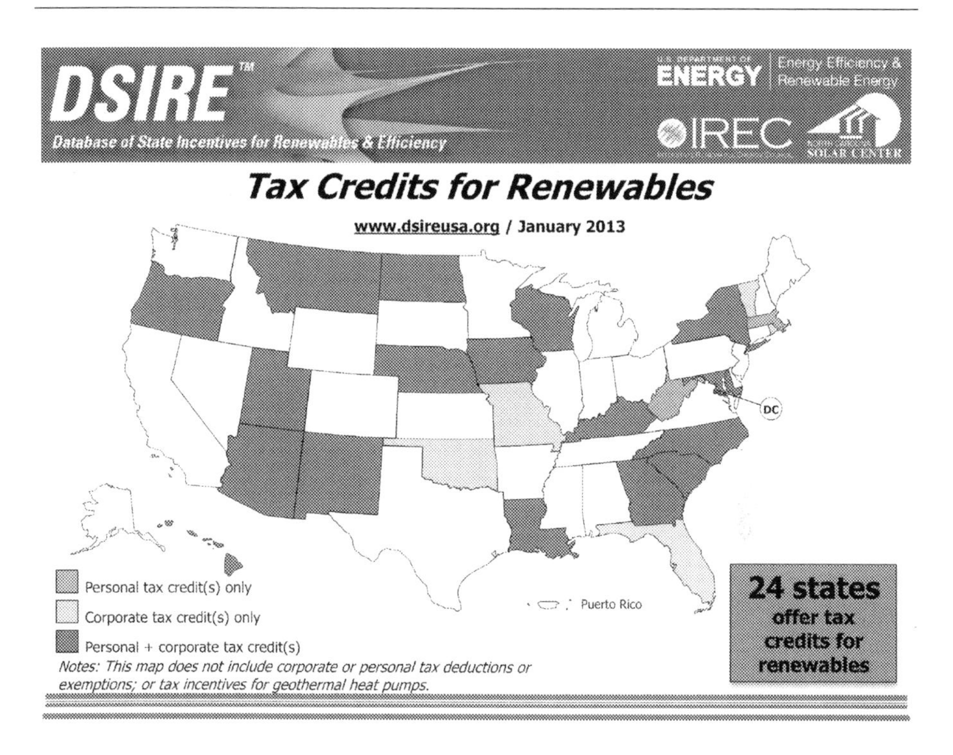

There is another, equally important aspect to tax credits, particular to larger projects and focused on investments. These commercial tax credits allow developers and third-party financiers to take advantage of reducing tax liability to defray project costs. According to North Carolina State University's analysis, "tax credits generally range from 10% to 50% of project costs [and the] maximum credit generally ranges from $500 to $35,000 for residential systems and from $25,000 to $60 million for commercial systems."[12]

While tax credits often cover solar-electric and solar thermal systems, they do have some important drawbacks, primarily in that they only function for entities with tax liabilities. As a result, many nonprofits, educational institutions, churches, and government agencies cannot benefit from them. Third-party financing and power purchase agreements (so-called "solar leases") have mitigated some of these concerns because the developer/installer can take the credit, but the reliance on tax credits has slowed the installation of solar systems in some segments of the market.

[12]http://www.dsireusa.org/solar/solarpolicyguide/?id=13.

As of June 2013, about 20 states are represented in DSIRE's database for tax incentives, included below:

Photovoltaic

State	Program Name	Eligible Recipients	Incentive Amount
Arizona	Non-Residential Solar & Wind Tax Credit (Corporate)	Any nonresidential installation is eligible, including those for nonprofits and governments. Individuals, corporations and S corporations, and partnerships may claim the credit. Third-party financiers/ installers/mfrs. of eligible system may claim the credit.	10%
Arizona	Non-Residential Solar & Wind Tax Credit (Personal)	Any nonresidential installation is eligible, including those for nonprofits and governments. Individuals, corporations and S corporations, and partnerships may claim the credit. Third-party financiers/ installers/mfrs. of eligible system may claim the credit.	10%
Arizona	Renewable Energy Production Tax Credit (Corporate)	Any AZ taxpayer with an eligible system installed on or after 12/31/2010 which transmits electricity to a public or private electric transmission or distribution utility system. Includes individuals, partners in a partnership, members of an LLC, and shareholders of an S corporation.	Varies by year, paid for 10 years
Arizona	Renewable Energy Production Tax Credit (Personal)	Any AZ taxpayer with an eligible system installed on or after 12/31/2010 which transmits electricity to a public or private electric transmission or distribution utility system. Includes individuals, partners in a partnership, members of an LLC, and shareholders of an S corporation.	Varies by year, paid for 10 years
Arizona	Residential Solar and Wind Energy Systems Tax Credit	Residents who are not a dependent of another taxpayer	25%

(*continued*)

Photovoltaic (Continued)

State	Program Name	Eligible Recipients	Incentive Amount
Florida	Renewable Energy Production Tax Credit	A nonresidential taxpayer with facility placed in service or expanded after May 1, 2012. The credit is for electricity produced and sold by the taxpayer to an unrelated party during a given tax year. Florida corporate income taxpayers that own an interest in a general partnership, limited partnership, limited liability company, trust, or other artificial entity that owns a Florida renewable energy facility can apply for this credit.	$0.01/kWh
Georgia	Clean Energy Tax Credit (Corporate)	Any GA taxpayer who has constructed, purchased, or leased renewable energy property and placed it in service.	35%
Georgia	Clean Energy Tax Credit (Personal)	Any GA taxpayer who has constructed, purchased, or leased renewable energy property and placed it in service.	35%
Hawaii	Solar and Wind Energy Credit (Corporate)	HI taxpayer that files a corporate net income tax return or franchise tax return. Credit may be claimed for every eligible renewable energy technology system that is installed and placed in service. Third-party taxpaying entities may claim the credit if they install and own a system on a commercial taxpayer's building or on a nonprofit or government building. Multiple owners of a single system may take a single tax credit. The credit is apportioned between the owners in proportion to their contribution to the system's cost.	35%

(continued)

Photovoltaic (Continued)

State	Program Name	Eligible Recipients	Incentive Amount
Hawaii	Solar and Wind Energy Credit (Personal)	HI taxpayer that files an individual net income tax return. Credit may be claimed for every eligible renewable energy technology system that is installed and placed in service. Credit may be claimed for every eligible renewable energy technology system that is installed and placed in service. Multiple owners of a single system may take a single tax credit. The credit is apportioned between the owners in proportion to their contribution to the system's cost.	35%
Iowa	Renewable Energy Production Tax Credit (Personal)	Producers or purchasers of renewable energy from qualified facilities. Installations must be at least 51% owned by a state resident or other qualifying owner, and placed in service on or after July 1, 2005, and before January 1, 2012. Electricity must be sold to an unrelated person to qualify for the tax credit.	$0.015/kWh for 10 years after energy production begins
Iowa	Renewable Energy Production Tax Credits (Corporate)	Producers or purchasers of renewable energy from qualified facilities. Installations must be at least 51% owned by a state resident or other qualifying owner, and placed in service on or after July 1, 2005, and before January 1, 2012. Electricity must be sold to an unrelated person to qualify for the tax credit.	$0.015/kWh for 10 years after energy production begins
Iowa	Solar Energy Systems Tax Credit (Corporate)		15%

(continued)

Photovoltaic (Continued)

State	Program Name	Eligible Recipients	Incentive Amount
Iowa	Solar Energy Systems Tax Credit (Personal)		15%
Kentucky	Renewable Energy Tax Credit (Corporate)	Any installation on a dwelling unit or on property that is owned and used by the taxpayer as commercial property.	$3.00/watt (DC)
Kentucky	Renewable Energy Tax Credit (Personal)	A system installed on a dwelling unit located in the state.	$3.00/watt (DC)
Kentucky	Tax Credits for Renewable Energy Facilities	Companies that build or renovate facilities that utilize renewable energy.	Negotiated on a case-by-case basis
Louisiana	Tax Credit for Solar and Wind Energy Systems on Residential Property (Corporate)	A taxpayer who purchases and installs an eligible system or who purchases a new home with such a system already in place.	50%
Louisiana	Tax Credit for Solar and Wind Energy Systems on Residential Property (Personal)	A taxpayer who purchases and installs an eligible system or who purchases a new home with such a system already in place.	50%
Maryland	Clean Energy Production Tax Credit (Corporate)	All individuals and corporations that sell electricity produced by a qualified facility to an unrelated person. Net metering arrangements qualify.	$0.0085/kWh for 5 years after facility is placed in service
Maryland	Clean Energy Production Tax Credit (Personal)	All individuals and corporations that sell electricity produced by a qualified facility to an unrelated person. Net metering arrangements qualify.	$0.0085/kWh for 5 years after facility is placed in service
Massachusetts	Residential Renewable Energy Income Tax Credit	Owner or tenant of residential property who occupies property as his or her principal residence.	15% of purchase price and installation cost less any federal tax credits or rebates received from the US HUD

(*continued*)

Photovoltaic (Continued)

State	Program Name	Eligible Recipients	Incentive Amount
Montana	Alternative Energy Investment Tax Credit (Corporate)	A corporation, partnership, or small business corporation which makes a minimum investment of $5000.	35%
Montana	Alternative Energy ITC (Personal)	An individual who makes a minimum investment of $5000.	35%; participant investment must be greater than or equal to $5000
Montana	Residential Alternative Energy System Tax Credit	MT resident individual who installs an eligible system in their principal dwelling.	100%
Nebraska	Renewable Energy Tax Credit (Corporate)	New renewable electric generation facility (in operation on or after July 14, 2006)	Credits are available for a 10-year period: $0.00075/kWh for electricity generated through 9/30/2007; $0.001/kWh from 10/1/2007 to 12/31/2009; $0.00075/kWh from 1/1/2010 to 12/31/2012; $0.0005/kWh on or after 1/1/2013
Nebraska	Renewable Energy Tax Credit (Personal)	New renewable electric generation facility (in operation on or after July 14, 2006)	Credits are available for a 10-year period: $0.00075/kWh for electricity generated through 9/30/2007; $0.001/kWh from 10/1/2007 to 12/31/2009; $0.00075/kWh from 1/1/2010 to 12/31/2012; $0.0005/kWh on or after 1/1/2013
New Mexico	Advanced Energy Tax Credit (Corporate)	Any taxpayer	6%
New Mexico	Advanced Energy Tax Credit (Personal)	Any taxpayer	6%

(continued)

Photovoltaic (Continued)

State	Program Name	Eligible Recipients	Incentive Amount
New Mexico	Renewable Energy Production Tax Credit (Corporate)	A taxpayer who holds title to a qualified energy generator that first produced electricity on or before January 1, 2018; or a taxpayer who leases property upon which a qualified energy generator operates from a county or municipality under authority of an industrial revenue bond and if the qualified energy generator first produced electricity on or before January 1, 2018.	Varies annually over 10 years; $0.027/kWh average
New Mexico	Renewable Energy Production Tax Credit (Personal)	A taxpayer who holds title to a qualified energy generator that first produced electricity on or before January 1, 2018; or a taxpayer who leases property upon which a qualified energy generator operates from a county or municipality under authority of an industrial revenue bond and if the qualified energy generator first produced electricity on or before January 1, 2018.	Varies annually over 10 years; $0.027/kWh average
New Mexico	Solar Market Development Tax Credit	Residents and noncorporate businesses, including agricultural enterprises.	10% of purchase and installation costs
New York	Residential Solar Tax Credit	Residential taxpayers that install eligible solar equipment at his or her principal residence.	25%
North Carolina	Renewable Energy Tax Credit (Corporate)	Any NC taxpayer who has constructed, purchased, or leased renewable energy property and placed it in service.	35% (distributed 7% per year for 5 years for nonresidential installations)
North Carolina	Renewable Energy Tax Credit (Personal)	Any NC taxpayer who has constructed, purchased, or leased eligible equipment and placed it in service.	35%

(*continued*)

Photovoltaic (Continued)

State	Program Name	Eligible Recipients	Incentive Amount
North Dakota	Renewable Energy Tax Credit	Corporate taxpayers filing a North Dakota income tax return. System must be installed on a building or on property owned or leased by the taxpayer in North Dakota.	15% (distributed 3% per year for 5 years)
Oklahoma	Zero-Emission Facilities Production Tax Credit	Any nonresidential taxpayer who sells electricity to an unrelated person. Any nontaxable entities, including agencies of the State of Oklahoma, may transfer their credit to a taxpayer.	$0.0050/kWh for first 10 years of operation
Oregon	Residential Energy Tax Credit	Installations on residential real property. Can be claimed by homeowners, renters, landlords, third parties.	$2.10/watt (DC) at standard test conditions (STC)
South Carolina	Solar Energy and Small Hydropower Tax Credit (Corporate)	Taxpayers who purchase and install an eligible system in or on a facility owned by the taxpayer.	25%
South Carolina	Solar Energy and Small Hydropower Tax Credit (Personal)	Any taxpayer who purchases and installs an eligible system in or on a facility owned by the taxpayer.	25% of eligible costs
Utah	Alternative Energy Development Incentive (Corporate)	An entity which conducts business within the state, including pass-through entities	75% of new state tax revenues (including, state, corporate, sales, and withholding taxes) over the life of the project or 20 years, whichever is less
Utah	Alternative Energy Development Incentive (Personal)	An entity which conducts business within the state, including pass-through entities	75% of new state tax revenues (including, state, corporate, sales, and withholding taxes) over the life of the project or 20 years, whichever is less

(*continued*)

Photovoltaic (Continued)

State	Program Name	Eligible Recipients	Incentive Amount
Utah	Renewable Energy Systems Tax Credit (Corporate)	Any company that owns a qualified system.	Residential: 25%; Commercial: 10%
Utah	Renewable Energy Systems Tax Credit (Personal)	Any Utah taxpayer who owns a qualified system.	Residential: 25%; Commercial: 10%
Vermont	Investment Tax Credit	Businesses that pay business income tax in Vermont that do not receive grants/funding from CEDF.	30% of expenditures (for systems placed into service on or before 09/01/2011). 7.2% for property placed in service on or after that date and on or before 12/31/2016; 2.4% after that date
West Virginia	Residential Solar Energy Tax Credit	Any taxpayer who installs (or contracts for the installation of) an eligible solar system on residential property that he or she owns and uses as a residence in the state.	30%

Solar Water Heating

State	Program Name		Incentive Amount
Arizona	Non-Residential Solar & Wind Tax Credit (Corporate)	Any nonresidential installation is eligible, including those for nonprofits and governments. Individuals, corporations and S corporations, and partnerships may claim the credit. Third-party financiers/installers/mfrs. of eligible system may claim the credit.	10%
Arizona	Non-Residential Solar & Wind Tax Credit (Personal)	Any nonresidential installation is eligible, including those for nonprofits and governments. Individuals, corporations and S corporations, and partnerships may claim the credit. Third-party financiers/installers/mfrs. of eligible system may claim the credit.	10%

(continued)

Solar Water Heating (Continued)

State	Program Name		Incentive Amount
Arizona	Residential Solar and Wind Energy Systems Tax Credit	Residents who are not a dependent of another taxpayer	25%
Georgia	Clean Energy Tax Credit (Corporate)	Any GA taxpayer who has constructed, purchased, or leased renewable energy property and placed it in service.	35%
Georgia	Clean Energy Tax Credit (Personal)	Any GA taxpayer who has constructed, purchased, or leased renewable energy property and placed it in service.	35%
Hawaii	Solar and Wind Energy Credit (Corporate)	Taxpayer that files a corporate net income tax return or franchise tax return. Credit may be claimed for every eligible renewable energy technology system that is installed and placed in service. Systems installed on new residential construction after December 31, 2009, are not eligible. Third-party taxpaying entities may claim the credit if they install and own a system on a commercial taxpayer's building or on a nonprofit or government building. Multiple owners of a single system may take a single tax credit. The credit is apportioned between the owners in proportion to their contribution to the system's cost.	35%
Hawaii	Solar and Wind Energy Credit (Personal)	Taxpayer that files an individual net income tax return. Credit may be claimed for every eligible renewable energy technology system that is installed and placed in service. Systems installed on new residential construction after December 31, 2009, are not eligible. Multiple owners of a single system may take a single tax credit. The credit is apportioned between the owners in proportion to their contribution to the system's cost.	35%

(continued)

Solar Water Heating (Continued)

State	Program Name		Incentive Amount
Iowa	Solar Energy Systems Tax Credit (Corporate)		15%
Kentucky	Renewable Energy Tax Credit (Corporate)	Any installation on a dwelling unit or on property that is owned and used by the taxpayer as commercial property.	30%
Kentucky	Renewable Energy Tax Credit (Personal)	A system installed on a dwelling unit located in the state.	30%
Louisiana	Tax Credit for Solar and Wind Energy Systems on Residential Property (Corporate)	A taxpayer who purchases and installs an eligible system or who purchases a new home with such a system already in place.	50%
Louisiana	Tax Credit for Solar and Wind Energy Systems on Residential Property (Personal)	A taxpayer who purchases and installs an eligible system or who purchases a new home with such a system already in place.	50%
Massachusetts	Residential Renewable Energy Income Tax Credit	Owner or tenant of residential property who occupies property as his or her principal residence.	15% of purchase price and installation costs, less any federal tax credits or rebates received from the US HUD
Montana	Residential Alternative Energy System Tax Credit	MT resident individual who installs the system in their principal dwelling.	100%
New Mexico	Solar Market Development Tax Credit	Residents and noncorporate businesses, including agricultural enterprises.	10% of purchase and installation costs, less any eligible federal tax credit.
New York	Residential Solar Tax Credit	Residential taxpayers who install eligible solar equipment at his or her principal residence.	25%
North Carolina	Renewable Energy Tax Credit (Corporate)	Any taxpayer who has constructed, purchased, or leased renewable energy property places it in service.	35% (distributed 7% per year for 5 years for nonresidential installations)

(continued)

Solar Water Heating (Continued)

State	Program Name		Incentive Amount
North Carolina	Renewable Energy Tax Credit (Personal)	Any taxpayer who has constructed, purchased, or leased renewable energy property places it in service.	35%
North Dakota	Renewable Energy Tax Credit	Corporate taxpayers filing a North Dakota income tax return. System must be installed on a building or on property owned or leased by the taxpayer in North Dakota.	15% (distributed 3% per year for 5 years)
Oregon	Residential Energy Tax Credit	Installations on residential real property. Can be claimed by homeowners, renters, landlords, and third parties.	$0.60/kW energy saved in first year
South Carolina	Solar Energy and Small Hydropower Tax Credit (Corporate)	Taxpayers who purchase and install an eligible system in or on a facility in South Carolina and owned by the taxpayer.	25%
South Carolina	Solar Energy and Small Hydropower Tax Credit (Personal)	Taxpayers who purchase and install an eligible system in or on a facility in South Carolina and owned by the taxpayer.	25% of eligible costs
Utah	Renewable Energy Systems Tax Credit (Corporate)	Any company that owns a qualified system.	Residential: 25%; Commercial: 10%
Utah	Renewable Energy Systems Tax Credit (Personal)	Any Utah taxpayer who owns a qualified system.	Residential: 25%; Commercial: 10%
Vermont	ITC	Business that pay business income tax in Vermont that do not receive grants/funding from CEDF.	30% of expenditures (for systems placed into service on or before 09/01/2011); 7.2% for property placed in service on or after that date and on or before 12/31/2016; 2.4% after that date

(continued)

Solar Water Heating (Continued)

State	Program Name		Incentive Amount
West Virginia	Residential Solar Energy Tax Credit	Any taxpayer who installs (or contracts for the installation of) an eligible solar system on residential property that he or she owns and uses as a residence in the state.	30%

Direct Cash Financing

This broad category of incentives includes rebates, grants, and performance-based incentives. The most simple form of direct cash financing is *a rebate*, which is simply a payment made to the purchaser after installation. This is similar to typical consumer rebates; however, because of the complexity and safety requirements for solar systems, several inspections are often required prior to installation. Rebates range in value. Massachusetts is an example of a very strong solar rebate, offering up to $2500 to purchasers of residential solar systems.

Similar to rebates, a *buydown* is a reduction in the bottom-line cost of a system to a buyer, where the funder makes a direct payment to the seller or installer as opposed to the consumer. This can be beneficial to consumers as it tends to reduce the potential for lag between going out of pocket and receiving the rebate.

Grants are generally not as common for residential installations and usually involve competitive applications. Solar program administrators and consumer often use rebates, buydowns, and grants interchangeably, as all three are typically based on system size, capital costs, or system performance.[13]

While some incentives, such as the Massachusetts solar rebate program, use expected system output as their basis for payment, but performance-based incentives, however, are paid by "actual energy output of a solar energy system (to encourage optimal system design and installation) and are disbursed over several years. FITs and renewable energy credit (REC) purchase programs are examples of performance-based incentives. Direct cash incentives may be offered by states, local governments and/or utilities."[14]

Direct cash incentives are Keynesian in nature, in that they serve to stimulate the market to encourage efficiency and investment throughout the value chain, ultimately dropping the price of solar products and services. Specifically, direct cash incentives offer the following benefits:

1. Reduced up-front costs (rebates, buydowns, and some grants, etc.);
2. Continued and supported revenue stream (performance-based incentives);
3. Reduced return on investment and lower overall costs to consumers;
4. Offered immediately (or nearly so) rather than at tax year-end and are provided regardless of tax liability.

[13] http://www.dsireusa.org/solar/solarpolicyguide/?id=10.
[14] *id.*

In essence, these programs distribute some of the costs not over time as Property Assessed Clean Energy (PACE) does, but across the populace. The cash to support these programs comes from taxpayers or rate payers, on the premise that cost distribution should accompany benefit distribution, as the climate, security, and peak production of solar benefits all electric customers and not just the end-purchaser.

As of mid-2013, over 20 states and 200 utilities offer some form of direct cash financing. Generally, these programs defray 10–30% of the cost of the system and are seen as being one of the most important components to competitive pricing and a burgeoning solar installation market in the United States.[15] According to the 2012 Solar Jobs Census, a lack of consumer incentives was the second most reported obstacle to growth, suggesting that even more can be done with direct cash incentives.[16]

Despite the popularity of these incentives, they have many detractors. In addition to legislators who do not generally believe in Keynesian Economic Theory or externalized cost distribution, direct cash incentives require direct funding mechanisms. These mechanisms are generally more difficult to develop, can be more costly to administer, and are often subject to raid when budgets become tight.[17]

International comparisons show the promise of a mechanism that is gaining popularity among US states called a FIT. This mechanism works by requiring utilities to purchase excess power from solar systems at a fixed rate for a specific number of years. Many programs also require the utilities to pay the renewable energy credits, if applicable. The FIT system is the basis of the European model, addressed later in this chapter, because it provides a stable, ongoing revenue stream and market for the renewable energy.

Recently, the North Carolina Solar Center, at NC State University, conducted an analysis of direct cash incentives that is a valuable summary of the benefits. An excerpt of that analysis follows (which has been republished with their permission):

Most state programs have since adopted more complex incentive structures to incorporate and address four primary issues that have emerged as solar markets have evolved:

- ***The different tax treatment of residential, commercial and non-profit sectors***. About one-third of state PV programs and several state solar water heating programs provide larger incentives to the government/non-profit sector because these entities cannot easily take advantage of state and federal tax credits.
- ***The desire to reward system performance rather than system capacity***. Performance-based incentives, which provide project owners with long-term payments based on electricity production on a dollar-per-kilowatt-hour basis, have gained increasing attention. So, too, have hybrid approaches, which sometimes involve upfront rebates based on a system's expected performance. Such incentives are based on system capacity but may be adjusted after taking into consideration certain other factors, including system rating, location, tilt, and orientation and shading. Payments based on performance or expected performance

[15] *id.*

[16] http://thesolarfoundation.org/research/national-solar-jobs-census-2012.

[17] http://www.dsireusa.org/solar/solarpolicyguide/?id=10.

rather than capital investment have gained prominence as a means of incentivizing proper system design and installation.

- ***Other mechanisms to protect consumers and guarantee adequate performance***. Ensuring that solar energy systems will perform as expected solidifies consumer confidence and helps guarantee that a state is making prudent investments. Beyond tying incentive payments to actual system performance, states have developed quality-assurance mechanisms that include one or more of the following provisions: equipment and installation standards; warranty requirements; installer requirements, assessments, and voluntary training; design standards and administrative design review; post-installation site inspections and acceptance testing; performance monitoring and assessment; and maintenance requirements and services. The best approach will ultimately depend on the performance issues of greatest concern and will differ depending on each program's objectives and constraints [3].
- ***Interest in rewarding high-value or emerging applications***. Providing bonus incentives for desirable applications is becoming increasingly common among state programs. Current and past examples include higher incentives for affordable housing; for the use of in-state manufactured components; for the use of building-integrated PV; for the use of certified installers, and installations in certified green buildings; and for solar on Energy Star homes, new construction and public buildings.[18]

Below are the DSIRE tables of state solar cash incentives for PV and water heating as of mid-2013:

PV

State	Program Name	Eligible Recipients	Incentive Amount
California	California Solar Initiative—Multi-Family Affordable Solar Housing (MASH) Program	Owners or operators of existing multifamily affordable housing that meets the definition of low-income residential housing in Pub. Util. Code_ 2852.8.	PV System Offsetting Common Area Load: $3.30/W AC; PV System Offsetting Tenant Load: $4.00/W AC; Incentives may be reduced based on expected performance
California	California Solar Initiative—PV Incentives	All customers of investor-owned and publicly owned California utilities	Varies by sector and system size
California	California Solar Initiative—Single-Family Affordable Solar Housing (SASH) Program	Must be a customer of PG&E, SCE, or SDG&E, and the household's total income must be 80% of the area median income (AMI) or less based on the most recent available income tax return.	Varies depending on participant's income level

(*continued*)

[18] *id.*

PV (Continued)

State	Program Name	Eligible Recipients	Incentive Amount
California	CEC—New Solar Homes Partnership	Home builders	Varies. There are separate levels for new custom homes and homes in small developments, homes that are a part of large developments, individual units of low-income housing, and common areas of low-income housing developments. Incentives are adjusted based on expected performance, and will decline over time based on the total installed capacity
Connecticut	Residential Solar Investment Program	Residential up to 10 kW.	Customer owned: First 5 kW \$1.75/W, next 5 kW \$0.55/W; third-party owned: \$0.300/kWh for 6 years
Delaware	Delmarva Power—Green Energy Program Incentives	All electric distribution customers of Delmarva Power and Light	Residential, nonresidential: \$1.25/W DC for first 5 kW, \$0.75/W for next 5 kW, \$0.35/W for next 40 kW; nonprofit: \$2.55/W DC for first 5 kW, \$1.50/W for next 5 kW, \$0.70/W for next 40 kW. PV system cost may not exceed \$12/W
District of Columbia	Renewable Energy Incentive Program	Pepco customers located within the District of Columbia. Multifamily homes, single family homes, businesses, nonprofits, institutional (excluding government), and private schools are eligible.	Solar PV: \$0.50/W DC
Illinois	Community Solar and Wind Grant Program	Businesses, government, and nonprofit customers of a utility that imposes the Renewable Energy Resources and Coal Technology Development Assistance Charge.	Business PV: \$1.50/W or 25% of project costs; government and nonprofit PV: \$2.60/W or 40% of project costs

(continued)

PV (Continued)

State	Program Name	Eligible Recipients	Incentive Amount
Illinois	Solar and Wind Energy Rebate Program	Customers of investor-owned electric or gas utilities; customers of electric cooperative or municipal utilities that impose the Renewable Energy Resources and Coal Technology Development Assistance Charge.	Residential and commercial solar PV: $1.50/W or 25% of project costs; public sector and nonprofit solar PV: $2.60/W or 40% of project costs
Maine	Efficiency Maine Renewable Energy Program	Owners and tenants, who are Maine residents, of residential or commercial properties located in Maine	$500 per 1000 kWh est. annual production
Maryland	Commercial Clean Energy Grant Program	In-state businesses, nonprofits, and local government	$60/kW for systems less than 100 kW; $30/kW for systems between 100 kW and 200 kW (*Note*: Ranges are mutually exclusive, not additive)
Maryland	Residential Clean Energy Grant Program	State residents; must be a primary residence	$1000
Massachusetts	Commonwealth Solar II Rebates	Eligibility includes all customers of IOUs, certain MLPs. Applicant does not have to be the future owner of PV project but must be the consumer of system-generated electricity.	$0.40/W DC base; $0.05/W DC adder for MA components; $0.40/W DC adder for moderate home value or for moderate income (moderate home value and moderate income applicable to residential rebates only)
Nevada	Renewable Generations Rebate Program	Existing grid-connected customer of NV Energy	Residential and small business: $1.35/W AC; public facilities/schools: $3.10/W AC
New Hampshire	Commercial & Industrial Solar Rebate Program	Nonresidential entities (including for profit, nonprofit, schools, government, and multifamily residential, as long as there is some common electricity/energy offset involved).	$0.80/W (DC) for new systems; $0.50/W (DC) for additions to existing systems

(*continued*)

PV (Continued)

State	Program Name	Eligible Recipients	Incentive Amount
New Hampshire	Renewable Energy Rebate Program	Any residential owner of a PV system that begins operation on or after 7/1/2008 and is located at the owner's residence.	$0.75/W DC (STC)
New York	PV Incentive Program	All customers who pay the RPS surcharge as part of their electric bill.	$1.40/W DC
Oregon	Solar Electric Incentive Program	Customers of Pacific Power and PGE	Residential (homeowner or third-party owned): $0.75/W DC for Pacific Power and PGE; nonresidential PGE, 0–35 kW: $1.20/W nonresidential Pacific Power, 0–35 kW: $1.10/W non-residential PGE, 35–200 kW: $0.60–$1.20/W nonresidential Pacific Power, 35–200 kW: $0.50–$1.10
Pennsylvania	Pennsylvania Sunshine Solar Rebate Program	Must be a state resident or an in-state small business with 100 or fewer full-time employees.	Residential: $0.75/W DC; Commercial: $0.75/W DC for first 10 kW, $0.50/W DC for next 90 kW; Low-Income: 35% of installed costs; Residential Battery Back-up: $0.35 amp-h
Puerto Rico	Puerto Rico—Green Energy Fund Tier I Incentive Program	Owners or operators of systems located at residential, commercial, and industrial properties.	If the installed cost ($/W) is less than the applicable reference cost ($/W) set by the Energy Affairs Administration: 40% of total project cost if the installed cost ($/W) is greater than the applicable reference cost ($/W) set by then Energy Affairs Administration: System size (W) multiplied by 40% multiplied by applicable reference cost ($/W)
Rhode Island	Commercial-Scale Renewable-Energy Grants		20% of project funding

(*continued*)

PV (Continued)

State	Program Name	Eligible Recipients	Incentive Amount
Rhode Island	Small-Scale Solar Grants	Eligible applicants are solar contractors, neighborhood associations, or another grouping mechanism approved by the RIEDC.	25% of project funding
Vermont	Small-Scale Renewable Energy Incentive Program	All sectors	For systems that generate at least 1000 kWh/year per kW of rated DC capacity installed: Residential PV: $0.45/W generating capacity (DC) up to 10 kW. An additional efficiency adder may be available: $0.15/W. Commercial PV: $0.40/W DC up to 10 kW, maximum incentive of $4000. Efficiency adder: $0.15/W. Low-income housing nonprofits, municipalities, public schools PV: $1.50/W DC up to 10 kW, Efficiency adder: $0.15/W. Systems that generate less than 1000 kWh/year per kW of rated DC capacity may be eligible for a lower, prorated incentive payment
Wisconsin	Renewable Energy Incentives	Residential customers of participating Wisconsin utilities	$600/kW DC

Solar Water Heating: State Incentives

State	Program Name	Eligible Recipients	Incentive Amount
California	California Solar Initiative—Low-Income Solar Water Heating Rebate Program	Low-income residential customers with natural gas water heaters	Single family low income: $25.64 per therm displaced; Multifamily low income: $19.23 per therm displaced

(*continued*)

(Continued)

State	Program Name	Eligible Recipients	Incentive Amount
California	California Solar Initiative—Solar Thermal Program	Single family and multifamily residential and all nonresidential customers of a participating utility.	Single Family Residential Systems that displace natural gas: $18.59 per estimated therm displaced; Single Family Residential Systems that displace electricity or propane: $0.54 per estimated kWh displaced; Commercial/Multifamily Systems that displace natural gas: $14.53 per estimated therm displaced; Commercial/Multifamily Systems that displace electricity or propane: $0.42 per estimated kWh displaced
Connecticut	Commercial Solar Thermal Incentive Program	Must be customer or CL&P or UI and install solar water heating project that provides 50–80% of domestic hot water needs.	Calculated: $70 multiplied by the SRCC "C" rating multiplied by the number of collectors multiplied by the shading factor
Connecticut	Solar Thermal Incentive Program	Must be customer or CL&P or UI and install solar water heating project that provides 50–80% of domestic hot water needs.	Calculated: $70 multiplied by the SRCC "C" rating multiplied by the number of collectors multiplied by the shading factor
Delaware	Delmarva Power—Green Energy Program Incentives	All electric distribution customers of Delmarva Power and Light	Residential, nonresidential: $1.00/annual kWh displaced; Nonprofit: $2.00/annual kWh displaced
District of Columbia	Renewable Energy Incentive Program	Pepco customers located within the District of Columbia. Multifamily homes, single family homes, businesses, nonprofits, and institutional (excluding government), private schools are eligible and third-party owners.	Solar Thermal: 20% of the installed cost

(*continued*)

(Continued)

State	Program Name	Eligible Recipients	Incentive Amount
Hawaii	Solar Water Heater Rebate	Residential retrofits eligible. Commercial incentives available on case-by-case basis. Available on all islands except Kauai.	Residential: $1000; Commercial: $250 per 12,000 Btu/h derated capacity
Illinois	Solar and Wind Energy Rebate Program	Customers of investor-owned electric or gas utilities; customers of electric cooperative or municipal utilities that imposes the Renewable Energy Resources and Coal Technology Development Assistance Charge.	30% of project costs
Maine	Efficiency Maine Renewable Energy Program	Owners and tenants, who are Maine residents, of residential or commercial properties located in Maine	$500 per 1000 kWh est. annual production
Maryland	Commercial Clean Energy Grant Program	In-state businesses, nonprofits, and local government	$20/ft^2 systems with a collector area of at least 50 ft^2 and less than 250 ft^2; $10/ft^2 for systems between 250 ft^2 and 1200 ft^2 (*Note*: Ranges are mutually exclusive, not additive)
Maryland	Residential Clean Energy Grant Program	State residents; must be a primary residence	$500
Massachusetts	Commonwealth Solar Hot Water Commercial Program	Commercial, municipal, or multifamily owners served by the state's IOUs or participating municipal utilities.	$45 × number of collectors × SRCC Rating (Private); $55 × number of collectors × *SRCC Rating (Public)
Massachusetts	Commonwealth Solar Hot Water Residential Program	Eligibility includes all customers of IOUs, certain MLPs.	$45 × # of collectors × SRCC rating in kBtu/panel/day (Category D Mildly Cloudy Day)

(*continued*)

(Continued)

State	Program Name	Eligible Recipients	Incentive Amount
New Hampshire	Commercial & Industrial Solar Rebate Program	Nonresidential entities (including for profit, nonprofit, schools, government, and multifamily residential, as long as there is some common electricity/energy offset involved).	\$0.12/modeled or rated kBtu/year for new systems with up to 15 collectors; \$0.07/modeled or rated kBtu/year for new systems with more than collectors; \$0.04/modeled kBtu/year for additions to existing systems
New Hampshire	Residential Solar Water Heating Rebates	Residential (must be primary residence).	Systems rated output 5.5–19.9 MMBTU: \$1500; systems rated output 20–29.9 MMBTU: \$1700; systems rated output 30 MMBTU or more: \$1900
New York	Solar Thermal Incentive Program	All customers who pay the RPS surcharge on their electric bill.	\$1.50/kWh displaced annually, for displacement of up to 80% of calculated existing thermal load
Oregon	Solar Water Heating Incentive Program	Customers of PGE or Pacific Power if displacing electric water heating, or NW Natural or Cascade Natural Gas if displacing gas water heating.	Incentives for solar water heating based on expected first-year savings; Residential hot water: \$0.40/kWh (PGE/Pacific Power), \$8.00/therm (NW Natural/Cascade). Commercial hot water: \$0.40/kWh (PGE/Pacific Power), \$8/therm (NW Natural/Cascade). Residential Pools: \$3/ft^2 collector area. Commercial Pools over 1000 ft^2: \$0.10/kWh (PGE/Pacific Power), \$1.50/therm (NW Natural/Cascade). Commercial Pools under 1000 ft^2: \$3/ft^2 collector area (PGE/Pacific Power), \$2.25/ft^2 collector area (NW Natural/Cascade)
Pennsylvania	Pennsylvania Sunshine Solar Rebate Program	Must be a state resident or an in-state small business with 100 or fewer full-time employees.	35% of installed costs

(continued)

(Continued)

State	Program Name	Eligible Recipients	Incentive Amount
Rhode Island	Small-Scale Solar Grants	Eligible applicants are solar contractors, neighborhood associations, or another grouping mechanism approved by the RIEDC.	25% of project funding
Vermont	Small-Scale Renewable Energy Incentive Program	All sectors	Residential Solar Hot Water: $1.50/100 British thermal unit/day (Btu/day) up to 200 kBtu/day, efficiency adder of $0.50/100 Btu/day. Commercial Solar Hot Water: $1.50/100 Btu/day up to 200 kBtu/day, Efficiency adder of $0.55/100Btu/day. Special Category Solar Hot Water: $3.00/100 Btu/day up to 1500 kBtu/day, efficiency adder of $0.55/100Btu/day. Solar Hot Water incentives are available for public swimming pools for Special Category customers only
Virgin Islands	Solar Water Heater Rebate Program		OG-100 Rated Solar Water Heater: up to $1000 OG-300 Rated Solar Water Heater: $1250
Wisconsin	Renewable Energy Incentives	All customers of participating Wisconsin utilities	$0.35/kWh or $6/therm

PACE

One of the most promising programs, PACE financing, has also been the most challenging to enact. PACE operates by providing financing that spreads out the cost of renewable energy and energy efficiency home and business upgrades over time, using the property as the collateral. Importantly, PACE programs follow the *property* rather than the owner. As a result, building owners need not consider the payback horizon in terms of when they might sell the property, but rather focus on the impact at sale (which is generally expected to be quite positive).

PACE operates differently from other financing related to property improvements. First, there is no personal liability outside of the property, similar to municipal property tax. Second, there are typically no up-front closing costs. PACE typically operates at the municipal level, whereby the local government provides the financing and adds the capital expense to existing property tax or a special assessment (sometimes through a municipally owned utility).

PACE financing offers many benefits. These include financing that is fixed in cost and spread out over a long period of time; interest rates based on the value and tax capacity of a home rather than the borrower's credit history; attachment of the loan to the property rather than the individual, making it transferable at sale; and sometimes favorable tax status, similar to deductions for local property taxes.[19]

PACE programs must be approved by local government, and in some instances, by states (if those states do not have home rule). While many states and municipalities have acted over the last several years, the initial programs (favorably for consumers) placed the special assessments ahead of mortgages in priority. This put the assessments on equal footing with property taxes and behind mortgages, creating a stir in the banking industry.

According to the DSIRE[20]:

> *Local PACE programs are currently operating in at least nine states (California, Connecticut, Florida, Maine, Michigan, Minnesota, Missouri, New York and Wisconsin) and the District of Columbia. In 2012, California launched a massive PACE program that, at its inception, allowed non-residential property owners in 126 cities and 14 counties to finance renewable energy, energy efficiency and water-efficiency projects. In other states, many residential PACE programs, especially those in states that have positioned PACE financing as the senior lien on a property, are currently on hold due to challenges created by the Federal Housing Finance Authority's (FHFA) stance on these programs. The FHFA's guidance directly impacts residential PACE programs (but not non-residential PACE programs) and effectively makes most residential PACE programs that assign the senior lien to PACE financing impossible to implement. However, some states, such as Florida and Hawaii, already had a structure in place to allow local governments to finance solar-energy systems through PACE programs. In response to the FHFA restrictions, a few states have enacted legislation that explicitly removes the senior lien provision in PACE programs, granting PACE financing a subordinate lien instead. The FHFA must issue a final rule by September 16, 2013.*

PACE funding is perhaps the most important consumer-related driver to impact the future of the solar industry. Its primary benefits include:

1. Up-front financing. This is a major obstacle for a majority of would-be solar customers;
2. Transferability. People in the United States are mobile and have concerns about lingering financial liabilities in case of a move;
3. Low risk, low rates. Because the program is based on property values and tax capacity and not credit rating, the cost of financing is relatively low.

[19] Lawrence Berkeley National Laboratory and Clean Energy States Alliance, 2008. Property Tax Assessments as a Finance Vehicle for Residential PV Installations. Lawrence Berkeley National Laboratory and Clean Energy States Alliance.

[20] http://www.dsireusa.org/solar/solarpolicyguide/?id=26.

These benefits are increasingly important in the United States because the demographic most likely to install solar, and to have the capital and income to make the investments, tend to be closer to retirement and moving than average. PACE is a way for baby boomers, many of whom have strongly favorable views of solar energy, to partake in the market without leveraging their retirement savings.

Below is a summary of state-level PACE programs in place as of mid-2013:

Arkansas
- Local Option—Property Assessed Clean Energy Financing

California
- Local Option—Municipal Energy Districts
- California Enterprise Development Authority—Statewide PACE Program
- CaliforniaFIRST
- City of Palm Desert—Energy Independence Program
- City of San Francisco—GreenFinanceSF
- Los Angeles County—Commercial PACE
- Sonoma County—Energy Independence Program
- Western Riverside Council of Governments—Home Energy Renovation Opportunity (HERO) Financing Program
- Western Riverside Council of Governments—Large Commercial PACE

Colorado
- Local Option—Improvement Districts for Energy Efficiency and Renewable Energy Improvements

Connecticut
- Local Option—Commercial PACE Financing
- Local Option—Residential Sustainable Energy Program

District of Columbia
- Property Assessed Clean Energy Financing

Florida
- Local Option—Special Districts
- Miami-Dade County—Voluntary Energy Efficiency and Renewable Energy Program

Georgia
- Local Option—Special Improvement Districts

Hawaii
- Local Option—Special Improvement Districts

Illinois
- Local Option—Contractual Assessments for Renewable Energy and/or Energy Efficiency

Louisiana
- Local Option—Sustainable Energy Financing Districts

Maine
- Local Option—Property Assessed Clean Energy
- Maine PACE Loans

Maryland
- Local Option—Clean Energy Loan Program

Massachusetts
- Local Option—Energy Revolving Loan Fund

Michigan
- City of Ann Arbor—PACE Financing
- Local Option—Property Assessed Clean Energy

Minnesota
- Local Option—Energy Improvement Financing Programs

Missouri
- Jefferson City—Property Assessed Clean Energy
- Local Option—Clean Energy Development Boards

Nevada
- Local Option—Special Improvement Districts

New Hampshire
- Local Option—Energy Efficiency & Clean Energy Districts

New Jersey
- Local Option—Property Assessed Clean Energy Financing

New Mexico
- Local Option—Renewable Energy Financing District/Solar Energy Improvement Special Assessments

New York
- Local Option—Municipal Sustainable Energy Programs
- Town of Babylon—Long Island Green Homes Program

North Carolina
- Local Option—Clean Energy Financing

Ohio
- Local Option—Special Energy Improvement Districts

Oklahoma
- Local Option—County Energy District Authority

Oregon
- Local Option—Local Improvement Districts

Texas
- Local Option—Contractual Assessments for Energy Efficient Improvements

Utah
- Local Option—Commercial PACE Financing

Vermont
- Local Option—Property Assessed Clean Energy

Virginia
- Local Option—Clean Energy Financing

Wisconsin
- Local Option—Energy-Efficiency Improvement Loans
- River Falls Municipal Utilities—Renewable Energy Finance Program

Wyoming
- Local Option—Energy Improvement Loan Program

Property Tax Incentives

Property tax incentives are important for the long-term health of the solar industry, but are not generally stimulating activity. Such incentives essentially exclude the value of the system from the property assessment, so that homeowners do not have to pay additional property tax for the improvement. This puts solar systems in a different category from other improvements (such as swimming pools, sheds, and so on) but it is more of a removal of a potential negative rather than an affirmative addition.

This is not to minimize the importance of property tax incentives. With all of the potential barriers to homeowners (by now, it should be apparent that a much greater proportion of US incentive programs are focused on smaller-scale systems than in other parts of the world), increased property taxes would be another challenge. As a result, more than 30 states have enabling legislation regarding property tax incentives.[21]

As of June 2013, this is the DSIRE listing of property tax incentives by state:

Alaska
- Local Option—Property Tax Exemption for Renewable Energy Systems

Arizona
- Energy Equipment Property Tax Exemption
- Property Tax Assessment for Renewable Energy Equipment

California
- Property Tax Exclusion for Solar Energy Systems

Colorado
- Local Option—Property Tax Exemption for Renewable Energy Systems
- Property Tax Exemption for Residential Renewable Energy Equipment
- Renewable Energy Property Tax Assessment

Connecticut
- Local Option—Property Tax Exemption for Renewable Energy Systems
- Property Tax Exemption for Renewable Energy Systems

District of Columbia
- Solar Energy System and Cogeneration System Personal Property Tax Credit

Hawaii
- City and County of Honolulu—Real Property Tax Exemption for Alternative Energy Improvements

Illinois
- Special Assessment for Solar Energy Systems

Indiana
- Renewable Energy Property Tax Exemption

Iowa
- Property Tax Exemption for Renewable Energy Systems

Kansas
- Renewable Energy Property Tax Exemption

Louisiana
- Solar Energy System Exemption

Maryland
- Anne Arundel County—High Performance Dwelling Property Tax Credit
- Anne Arundel County—Solar and Geothermal Equipment Property Tax Credits
- Baltimore County—Property Tax Credit for High Performance Buildings and Homes
- Baltimore County—Property Tax Credit for Solar and Geothermal Devices
- Carroll County—Green Building Property Tax Credit
- Harford County—Property Tax Credit for Solar and Geothermal Devices

[21] http://www.dsireusa.org/solar/solarpolicyguide/?id=11.

- Howard County—High Performance and Green Building Property Tax Credit
- Local Option—Property Tax Credit for High Performance Buildings
- Local Option—Property Tax Credit for Renewables and Energy Conservation Devices
- Montgomery County—High Performance Building Property Tax Credit
- Prince George's County—Solar and Geothermal Residential Property Tax Credit
- Property Tax Exemption for Solar and Wind Energy Systems
- Special Property Assessment for Renewable Heating & Cooling Systems

Massachusetts
- Renewable Energy Property Tax Exemption

Michigan
- Alternative Energy Personal Property Tax Exemption

Minnesota
- Wind and Solar-Electric (PV) Systems Exemption

Missouri
- Renewable Energy Generation Zone Property Tax Abatement

Montana
- Corporate Property Tax Reduction for New/Expanded Generating Facilities
- Generation Facility Corporate Tax Exemption
- Renewable Energy Systems Exemption

Nevada
- Large-Scale Renewable Energy Property Tax Abatement (Nevada State Office of Energy)
- Property Tax Abatement for Green Buildings
- Renewable Energy Systems Property Tax Exemption

New Hampshire
- Local Option—Property Tax Exemption for Renewable Energy

New Jersey
- Assessment of Farmland Hosting Renewable Energy Systems
- Property Tax Exemption for Renewable Energy Systems

New Mexico
- Property Tax Exemption for Residential Solar Systems

New York
- Energy Conservation Improvements Property Tax Exemption
- Local Option—Real Property Tax Exemption for Green Buildings
- Local Option—Solar, Wind & Biomass Energy Systems Exemption
- New York City—Property Tax Abatement for Photovoltaic (PV) Equipment Expenditures

North Carolina
- Active Solar Heating and Cooling Systems Exemption
- Property Tax Abatement for Solar Electric Systems

North Dakota
- Renewable Energy Property Tax Exemption

Ohio
- City of Cincinnati—Property Tax Abatement for Green Buildings
- City of Cleveland—Residential Property Tax Abatement for Green Buildings
- Qualified Energy Property Tax Exemption for Projects 250kW or Less
- Qualified Energy Property Tax Exemption for Projects over 250kW (Payment in Lieu)

Oregon
- Local Option—Rural Renewable Energy Development Zones
- Renewable Energy Systems Exemption

Puerto Rico
- Puerto Rico—Property Tax Exemption for Solar and Renewable Energy Equipment

Rhode Island
- Local Option—Property Tax Exemption for Renewable Energy Systems
- Residential Solar Property Tax Exemption

South Dakota
- Renewable Energy System Exemption

Tennessee
- Green Energy Property Tax Assessment

Texas
- Renewable Energy Systems Property Tax Exemption

Vermont
- Local Option—Property Tax Exemption
- Uniform Capacity Tax and Exemption for Solar

Virginia
- Local Option—Property Tax Exemption for Solar

Wisconsin
- Solar and Wind Energy Equipment Exemption

Economic Development Incentives

Economic developers use various strategies to entice businesses to their local communities. These programs, which are very popular in more managed economies such as in China, typically offer tax incentives, loans, or other enticements to build production facilities or locate R&D centers. They are the hallmark of traditional economic development policy and are typically measured by their ability to create local jobs.

Such programs do have a downside, as seen in the highly publicized Solyndra bankruptcy. Often, economic development incentives create a race to the bottom, with intense competition among regions. Also, the impacts can be short term. Perhaps, the most highly publicized case of an economic incentive backfire was in the City of New London, which took property from residents by eminent domain (an act it had to defend all the way to the US Supreme Court) and offered property tax abatements to pharmaceutical giant Pfizer. Eight short years later, Pfizer left, leaving an enormous empty campus, 1400 lost jobs, and virtually nothing for the city coffers.

However, economic development policies that are implemented correctly and thoughtfully can have an impact, though policy makers must be sure that they are not in a race to the bottom, and companies must ensure that they are deal sweeteners and not necessary for firm survival. Often, these programs will require a minimum level of leveraged investment, commitment of funders, or job creation. One useful example of a thoughtful economic development policy is to guarantee a certain amount of purchased product (for a manufacturing facility) or agreement to buy a certain amount of power (from a project developer), which works particularly well when the municipality runs the electric utility.[22]

Many of the same players offer economic development strategies among US states. The DSIRE database shows about 20 programs in 2013:

Arizona
- Renewable Energy Business Tax Incentives

Arkansas
- Wind Energy Manufacturing Tax Incentive

California
- Sales and Use Tax Exclusion for Advanced Transportation and Alternative Energy Manufacturing Program

Connecticut
- Operational Demonstration Program
- Sales and Use Taxes for Items Used in Renewable Energy Industries

Federal
- Qualifying Advanced Energy Manufacturing Investment Tax Credit

Florida
- Miami-Dade County—Targeted Jobs Incentive Fund

Kansas
- Solar and Wind Manufacturing Incentive

Kentucky
- Incentives for Energy Independence

Massachusetts
- Alternative Energy and Energy Conservation Patent Exemption (Corporate)
- Alternative Energy and Energy Conservation Patent Exemption (Personal)

Michigan
- Energy Revolving Loan Fund—Clean Energy Advanced Manufacturing
- Nonrefundable Business Activity Tax Credit
- Refundable Payroll Tax Credit
- Renewable Energy Renaissance Zones

Mississippi
- Mississippi Clean Energy Initiative

Montana
- Alternative Energy Investment Tax Credit
- Property Tax Abatement for Production and Manufacturing Facilities

New Jersey
- Edison Innovation Clean Energy Manufacturing Fund—Grants and Loans
- Edison Innovation Green Growth Fund Loans
- Wind Manufacturing Tax Credit

New Mexico
- Alternative Energy Product Manufacturers Tax Credit

North Carolina
- Renewable Energy Equipment Manufacturer Tax Credit

Ohio
- Advanced Energy Job Stimulus Program

[22] http://www.dsireusa.org/solar/solarpolicyguide/?id=14.

Oklahoma
- Tax Credit for Manufacturers of Small Wind Turbines

Oregon
- Tax Credit for Renewable Energy Equipment Manufacturers

Pennsylvania
- Alternative and Clean Energy Program
- Solar Energy Incentives Program
- Wind and Geothermal Incentives Program

Puerto Rico
- Puerto Rico—Economic Development Incentives for Renewables

South Carolina
- Renewable Energy Manufacturing Tax Credit

Tennessee
- Green Energy Tax Credit
- Sales and Use Tax Credit for Emerging Clean Energy Industry

Texas
- Solar and Wind Energy Business Franchise Tax Exemption

Utah
- Alternative Energy Manufacturing Tax Credit

Virginia
- Clean Energy Manufacturing Incentive Grant Program
- Green Jobs Tax Credit
- Solar Manufacturing Incentive Grant (SMIG) Program

Washington
- Tax Abatement for Solar Manufacturers

Permitting

A common complaint of all businesses is the process for and cost of obtaining state and local permits. These permits for building, inspection, and design review are costly and can take a long time to complete. As a result, many jurisdictions have offered expedited and low- or no-fee permitting for solar projects.

North Carolina State University conducted a survey of various permitting plans across the country. They found that the cost of the permits ranged from $0 to $1200. Costs often rise as the system size increases.[23]

In addition to these hard costs, permits can cause substantial delays to solar deployment, typically because inspectors and other government agents do not understand solar systems. Expedited permitting, which literally puts the application to the "top of the stack," is a critical component to illustrating commitment to solar.[24]

Interestingly, however, when asked to report the major obstacles facing the growth of their businesses, solar employers across the value chain did not mention permitting (nor labor costs or other typical economic development favorites). General economic conditions, lack of incentives, and low customer awareness were the most common responses.[25]

[23] http://www.dsireusa.org/solar/solarpolicyguide/?id=16.

[24] *id.*

[25] The Solar Foundation 2012 Solar Jobs Census.

Because so many of these programs are offered at the local level, it is difficult to collect information on how many such programs exist. DSIRE includes the following programs on its list as of July 2013:

Arizona
- Maricopa Assn. of Governments—PV and Solar Domestic Water Heating Permitting Standards
- Maricopa County—Renewable Energy Systems Zoning Ordinance
- Solar Construction Permitting Standards

California
- Solar Construction Permitting Standards
- City of San Jose—Photovoltaic Permit Requirements
- San Diego County—Solar Regulations
- Santa Clara County—Zoning Ordinance

Colorado
- City and County of Denver—Solar Panel Permitting
- Solar Construction Permitting Standards

Connecticut
- Local Option—Building Permit Fee Waivers for Renewable Energy Projects

Florida
- Broward County Online Solar Permitting

Illinois
- Statewide Renewable Energy Setback Standards

Maryland
- Permits and Variances for Solar Panels, Calculation of Impervious Cover

Massachusetts
- Model As-of Right Zoning Ordinance or Bylaw: Allowing Use of Large-Scale Solar Energy Facilities

Nevada
- Clark County—Solar and Wind Building Permit Guides

New Jersey
- Solar and Wind Permitting Laws

Oregon
- City of Portland—Streamlined Building Permits for Residential Solar Systems
- Model Ordinance for Renewable Energy Projects
- Solar Permitting Law

Vermont
- Expedited Permitting Process for Solar Photovoltaic Systems

Virginia
- Guidelines for Solar and Wind Local Ordinances

Loan Programs

While loans have no impact on the overall cost of projects, they do dramatically reduce the up-front cost of any size solar project. At the residential level, the impact of this can be dramatic, because up-front costs can be significant; however, the payback is in reduced electrical bills over time. Even in some of the most competitive states in the US, this payback time can be too much for many consumers.

At the commercial level, loans allow for more projects and distributing fixed costs into the future, to align with cash flow. As with any investment, this allows capital to be used for expansion rather than being sunk into one project at a time.

States and municipalities can play a critical role in these loan programs by fully or partially guaranteeing the principal. Loan programs tend to be more politically palpable, and, if risks are managed properly, can result in revolving funds to spur more investment or actually produce a profit.[26]

These benefits seem compelling but are often not enough to encourage significant activity. As noted in Chapter 3, a relatively small percentage of consumers are willing to purchase "green" even if it costs more, and loans do nothing to reduce total costs. Despite this, DSIRE lists about 30 states that offer loan programs that include solar energy, summarized below:

Alabama
- AlabamaSAVES Revolving Loan Program
- Local Government Energy Loan Program
- South Alabama Electric Cooperative—Residential Energy Efficiency Loan Program
- TVA Partner Utilities—Energy Right Heat Pump Program

Alaska
- Power Project Loan Fund
- Small Building Material Loan

Arizona
- Sulphur Springs Valley EC—SunWatts Loan Program

Arkansas
- First Electric Cooperative—Home Improvement Loans

California
- Energy Efficiency Financing for Public Sector Projects
- SMUD—Residential Solar Loan Program

Colorado
- Boulder County—Elevations Energy Loans Program
- City and County of Denver—Elevations Energy Loans Program
- Direct Lending Revolving Loan Program
- Fort Collins Utilities—Residential On-Bill Financing Program

Connecticut
- Combined Heat and Power Pilot Loan Program
- CT Solar Loan
- Energy Conservation Loan
- Energy Efficiency Fund (Electric and Gas)—Residential Energy Efficiency Financing
- Low-Interest Loans for Customer-Side Distributed Resources

Federal
- Clean Renewable Energy Bonds (CREBs)
- Energy-Efficient Mortgages
- Qualified Energy Conservation Bonds (QECBs)
- U.S. Department of Energy—Loan Guarantee Program
- USDA—Biorefinery Assistance Program
- USDA—Rural Energy for America Program (REAP) Loan Guarantees

[26] http://www.dsireusa.org/solar/solarpolicyguide/?id=15.

Florida

- City of Lauderhill—Revolving Loan Program
- City of Tallahassee Utilities—Efficiency Loans
- City of Tallahassee Utilities—Solar Loans
- Clay Electric Cooperative, Inc—Energy Conservation Loans
- Clay Electric Cooperative, Inc—Solar Thermal Loans
- Gainesville Regional Utilities—Low-Interest Energy Efficiency Loan Program
- Orlando Utilities Commission—Residential Solar Loan Program
- St. Lucie County—Solar and Energy Loan Fund (SELF)

Georgia

- Athens-Clarke County—Green Business Revolving Loan Fund
- Flint Energies—Residential Energy Efficiency Loan Program
- Georgia Cities Foundation—Green Communities Revolving Loan Fund
- TVA Partner Utilities—Energy Right Heat Pump Program

Hawaii

- City and County of Honolulu—Solar Loan Program
- Farm and Aquaculture Alternative Energy Loan
- GreenSun Hawaii
- KIUC—Solar Water Heating Loan Program
- Maui County—Solar Roofs Initiative Loan Program

Idaho

- Idaho Falls Power—Energy Efficient Heat Pump Loan Program
- Low-Interest Energy Loan Programs

Illinois

- Green Energy Loans

Indiana

- South Central Indiana REMC—Residential Energy Efficiency Loan Program

Iowa

- Alliant Energy Interstate Power and Light (Gas and Electric)—Low Interest Energy Efficiency Loan Program
- Alternate Energy Revolving Loan Program
- IADG Energy Bank Revolving Loan Program
- Iowa Energy Bank
- MidAmerican Energy (Gas and Electric)—Residential EnergyAdvantage Loan Program

Kentucky

- Energy Efficiency Loans for State Government Agencies
- Energy Efficient Home Improvements Loan Program
- Greater Cincinnati Energy Alliance—Residential Loan Program
- Inter-County Energy Efficiency Loan Program
- Mountain Association for Community Economic Development—Energy Efficient Enterprise Loan Program
- Mountain Association for Community Economic Development—Solar Water Heater Loan Program
- TVA Partner Utilities—Energy Right Heat Pump Program

Louisiana

- City of Shreveport – Shreveport Energy Efficiency Program (SEED)
- Home Energy Loan Program (HELP)

Maine

- Seacoast Energy Initiative—Energy Efficiency Loan Program

Maryland
- Be SMART Multi-Family Efficiency Loan Program
- Jane E. Lawton Conservation Loan Program
- State Agency Loan Program

Massachusetts
- MassSAVE—HEAT Loan Program

Michigan
- City of Detroit—SmartBuildings Detroit Green Fund Loan
- Energy Revolving Loan Fund—Farm Energy
- Energy Revolving Loan Fund—Passive Solar
- Energy Revolving Loan Fund—Public Entities
- Michigan Saves—Business Energy Financing
- Michigan Saves—Home Energy Loan Program

Minnesota
- Agricultural Improvement Loan Program
- Fix-Up Loan
- Methane Digester Loan Program
- Minnesota Valley Electric Cooperative -Residential Energy Resource Conservation Loan Program
- Otter Tail Power Company—DollarSmart Energy Efficiency Loan Program
- Sustainable Agriculture Loan Program
- Value-Added Stock Loan Participation Program

Mississippi
- Energy Investment Loan Program
- Mississippi Power (Electric)—EarthCents Financing Program
- TVA Partner Utilities—Energy Right Heat Pump Program

Missouri
- Columbia Water & Light—Commercial Super Saver Loans
- Columbia Water & Light—Residential Super Saver Loans
- Energy Revolving Fund Loans

Montana
- Alternative Energy Revolving Loan Program

Nebraska
- Dollar and Energy Savings Loans

Nevada
- Revolving Loan Program
- Valley Electric Association—Solar Water Heating Program

New Hampshire
- Enterprise Energy Fund Loans
- Municipal Energy Reduction Fund

New Jersey
- Home Performance with Energy Star Program

New Mexico
- Drinking Water State Revolving Loan Fund

New York
- Home Performance with Energy Star Financing
- Residential Loan Fund

North Carolina
- Local Option—Financing Program for Renewable Energy and Energy Efficiency
- Lumbee River EMC—Solar Water Heating Loan Program

- Piedmont EMC—Residential Solar Loan Program
- Town of Carrboro—Worthwhile Investments Save Energy (WISE) Homes and Buildings Program
- TVA Partner Utilities—Energy Right Heat Pump Program

North Dakota

- Northern Plains EC—Residential and Commercial Energy Efficiency Loan Program
- Otter Tail Power Company—Dollar Smart Financing Program

Ohio

- Butler Rural Electric Cooperative—Energy Efficiency Improvement Loan Program
- Energy Conservation for Ohioans (ECO-Link) Program
- Energy Loan Fund
- Greater Cincinnati Energy Alliance—Residential Loan Program
- Hamilton County—Home Improvement Program

Oklahoma

- Community Energy Education Management Program
- Energy Loan Fund for Schools
- Higher Education Energy Loan Program
- Oklahoma Municipal Power Authority—WISE Energy Efficiency Loan Program
- Red River Valley REA—Heat Pump Loan Program

Oregon

- Ashland Electric Utility—Bright Way to Heat Water Loan
- Ashland Electric Utility—Residential Energy Efficiency Loan Program
- EWEB—Energy Management Services Loan
- EWEB—Residential Energy Efficiency Loan Programs
- EWEB—Residential Solar Water Heating Loan Program
- GreenStreet Lending Program
- Lane Electric Cooperative—Residential Energy Efficiency Loan Programs
- Salem Electric—Low-Interest Loan Program
- Small-Scale Energy Loan Program

Pennsylvania

- Adams Electric Cooperative—Energy Efficiency Loan Program
- Alternative and Clean Energy Program
- Energy Efficiency Loan Program
- High Performance Buildings Incentive Program
- Metropolitan Edison Company SEF Loans (FirstEnergy Territory)
- Penelec SEF of the Community Foundation for the Alleghenies Loan Program (FirstEnergy Territory)
- Small Business Pollution Prevention Assistance Account Loan Program
- Solar Energy Incentives Program
- Sustainable Development Fund Financing Program (PECO Territory)
- Sustainable Energy Fund (SEF) Loan Program (PPL Territory)
- West Penn Power SEF Commercial Loan Program
- Wind and Geothermal Incentives Program

South Carolina

- Berkeley Electric Cooperative—HomeAdvantage Efficiency Loan Program
- Blue Ridge Electric Cooperative—Heat Pump Loan Program
- ConserFund Loan Program
- Pee Dee Electric Cooperative—Energy Efficient Home Improvement Loan Program
- Santee Cooper—Renewable Energy Resource Loans

South Dakota
- Energy Efficiency Revolving Loan Program
- Otter Tail Power Company—Dollar Smart Financing Program
- Southeastern Electric—Electric Equipment Loan Program

Tennessee
- Commercial Energy Efficiency Loan Program
- Energy Efficient Schools Initiative—Loans
- TVA Partner Utilities—Energy Right Heat Pump Program

Texas
- Austin Energy—Residential Solar Loan Program
- City of Plano—Smart Energy Loan Program
- LoanSTAR Revolving Loan Program

Virginia
- Energy Project and Equipment Financing
- TVA Partner Utilities—Energy Right Heat Pump Program

Washington
- City of Seattle—Community Power Works Loan Program
- Clallam County PUD—Residential and Small Business Efficiency Loan Program
- Clallam County PUD—Residential and Small Business Solar Loan Program
- Clark Public Utilities—Residential Heat Pump Loan Program
- Clark Public Utilities—Solar Energy Equipment Loan
- Grays Harbor PUD—Residential Energy Efficiency Loan Program
- Grays Harbor PUD—Solar Water Heater Loan
- Port Angeles Public Works & Utilities—Solar Energy Loan Program
- Richland Energy Services—Residential Energy Conservation & Solar Loan Program
- Snohomish County PUD No 1—Solar Express Loan Program

Wisconsin
- City of Milwaukee—Milwaukee Shines Solar Financing

7 The Solar Labor Market—Efficiencies and Productivity

As discussed throughout this text, there are numerous variable costs that impact solar price points, including the type and cost of raw materials, capital and financing, transportation and fuel, and government tariffs and incentives. One of the most critical of these, particularly in the installation of solar systems, is labor. At the same "solar jobs" are an important global currency because politicians and policy makers seek to maximize the employment of their respective citizenries, and job creation is often a key component to the decision-making process.

In fact, measurements of solar employment garner major media attention in the United States. The health of the industry is often discussed by the measure of its total employment. The bigger, the better, seems to be the mantra of the US solar industry.

This philosophy, while paying short-term political dividends, is not amenable to long-term success in any industry. The US solar industry, still attempting to rid itself of its past perception as a fringe technology, uses solar employment figures to demonstrate how large and important it is. Soon, however, it must change gears and recognize that in the 21st century, industries win by being leaner and more efficient.

This does not mean that we should expect no, or even slow, growth in the solar industry. However, advances of 10–20% per year should not be considered the norm. As one solar executive put it, "we expect to increase our domestic installations by 50% year over year… if we increase our total employment by 20%, I'd say that is too much."

As with any labor market, the key to understanding the future (as best as it can be accomplished) is to analyze productivity. Productivity essentially refers to the amount of output per worker, which runs from zero to a theoretical maximum, which is normalized to 1.0. As productivity approaches the theoretical maximum, companies can no longer squeeze more work out of their existing employees, so they turn to new hires or turn away business. In the United States, companies are typically loath to turn away work, so generally prefer to pick up hiring as productivity gains momentum.

Labor productivity, had it been widely utilized, could have been used to predict the so-called jobless recovery in the United States following the global recession of 2008. According to the Bureau of Economic Analysis and the Bureau of Labor Statistics, GDP per worker in the United States in 2011 was $106,541, up 5% from 2007 and an increase of 16% from 2000.[1] While this seems dramatic, it represents a slower than historical rate (2.6% vs. 1.6% growth), as from 1960 to 2011, US GDP per worker increased 127% ($59,590).[2] This slower productivity growth rate means that there is

[1] http://www.bls.gov/ilc/intl_gdp_capita_gdp_hour.htm#table02.
[2] http://www.bls.gov/ilc/intl_gdp_capita_gdp_hour.htm#table02.

Solar Energy Markets. DOI: http://dx.doi.org/10.1016/B978-0-12-397174-6.00007-6

significant potential productivity to be gained, which means employers can squeeze more production out of their existing employees rather than hire new ones.

While productivity is a nearly failproof leading indicator of employment growth, its use requires more analysis in an emerging field like solar. Also, as it is a measurement of work performed by output, each segment must be analyzed separately, as the total gigawatt installed is different from the total number of pieces produced. This chapter is therefore divided by country and then value chain segment. Chapter 9 focuses on employers' needs and specific issues regarding the types of workers in the industry.

United States

The US solar industry is adding jobs at an impressive clip and has many of the hallmarks of a dynamic and innovative industry. The sector provides employment opportunities for nearly 120,000 workers at nearly 15,000 locations in all 50 states. And these jobs pay well, generally higher than the average wages offered by other related industries.

The solar energy workforce grew by 13.2% between 2011 and 2012, which is about six times the overall national employment growth rate over the same period. The 2012 growth was a capstone year as part of a 27% growth rate from 2010. This compares to only 3.2% national employment growth over that same period (August 2010–September 2012).

Solar employers remain optimistic despite some apparent challenges. In 2012, employers reported an anticipated growth rate of 17%. This figure does not even account for entrepreneurial activity.

Viewing job growth in the solar industry in relation to employment activity in conventional energy industries demonstrates just how impressive the growth figures have been. The fossil-fuel electric generation industry shed 3.77% of their workforce between 2011 and 2012 (a decline of 3857 jobs), while the coal mining labor force shrank by 0.83% (a loss of 851 workers).

Solar also outpaced growing fossil industries in 2011–2012. For example, the bulk power transmission and electric power distribution sectors both added workers (415 and 11,196 jobs, respectively) though at a much lower growth rate than solar (bulk power jobs grew 1.71% and electric power distribution by 5.13%).

As previously reported, the price of solar products is spurring an installation boom in the United States. Employers in the rapidly growing installation sector and interviewed for The Solar Foundation's 2012 Solar Jobs Census reported that low-priced PV is the key reason for their success.

In the second quarter of 2012, the average selling price for crystalline silicon PV modules was 44% less than in the same quarter in the previous year. Between the second quarter of 2011 and the second quarter of 2012, the average installed cost of solar fell by 33%. This has led to a large spike in installations.

If the trends and policies required to reduce the nonhardware costs of solar remain in place, it is likely to expect that installed costs will eventually fall as low as $1/W. This magic number, as seen in Chapter 9, is the "holy grail" at which electricity generated from solar in the United States will be cost competitive with wholesale electricity derived from conventional sources. Nearing this price point will drive growth dramatically in both installed solar capacity and new solar employment.

Price declines are fuel for the installers' fire, but are squeezing the majority of manufacturers. Some domestic manufacturers have found it difficult to keep up with the rapidly falling cost of solar components, leading some to scale back their operations or exit solar altogether. While the news is mostly bad for manufacturers, there are some brighter spots, particularly in thin film.

In addition to potential strength in thin film, it appears that while cell and module manufacturers are facing continued troubles, component and materials manufacturers (producing inverters, backsheets, racking, and metal paste) are holding their own and at least not shedding jobs. Reaching greater economies of scale is critical to manufacturing success (something that is currently more difficult in the United States than in other countries perhaps due to significantly lower public and private investment in manufacturing process and new facilities, as well as to present levels of domestic product demand).

US manufacturers have remained optimistic that global demand and more level playing fields, particularly if new trade agreements are struck, will benefit them in the long run. As noted in the Census, "if there are positive signs of long-term domestic demand, capital expenditures in US solar manufacturing may be justified." This is why, as noted throughout the report and particularly in Chapters 5 and 6, demand-side policies are so critical. The policies make solar more affordable while encouraging investment in manufacturing that allow production to run at a higher rate (thereby delivering production cost savings).[3]

Over the past few decades, the solar industry has grown from a niche market to a mainstream energy source. When polled, 9 out of 10 voters in the United States agreed it is important for the United States to develop and use solar power and approximately 70% said the federal government should do more than it currently is doing to promote solar.

The US solar industry includes nearly 15,000 solar establishments employing 119,016 solar workers as of summer 2012. Interestingly, year over year trends indicate a contraction in the number of solar businesses, but an employment growth rate of 13.2% from 2011 to 2012. This finding ultimately means that successful segments of the industry are growing rapidly, while others are closing shop or refocusing their businesses elsewhere.

In light of impressive employment growth trends, the decline in company count suggests consolidation and maturation of the US solar industry. For the last several years, solar employment has dramatically outpaced overall job growth (10 times faster in 2011 and 6 times faster in 2012), and many sectors across the various conventional energy generation industries either grew more slowly or experienced employment declines.

Solar compares quite favorably to the fossil-fuel electric generation sector, which lost nearly 4000 workers (representing a 3.77% decline in employment). At the same time, strong installed capacity figures and continued profitability in that sector are

[3]A lack of economies of scale is reported to be a reason why US solar manufacturing has fallen behind its global competitors.

fueling projections of growth by more than 17% in 2013. This represents approximately 20,000 new solar workers across the US value chain of solar activities.

As further evidence of US solar employer optimism, 44.2% of surveyed firms expected to add solar employees from summer 2012 to summer 2013, while only 3.6% expected to contract over the period. These findings show an industry that is growing at a much faster pace than the economy as a whole, which projects to add jobs at a rate of only 1.5% from 2012 to 2013.

The US solar industry is still heavily focused on PV, though solar water heating also continues to show strong growth. The following figure illustrates the breakdown of firms across the value chain by technology, as reported in The Solar Foundation Census.[4]

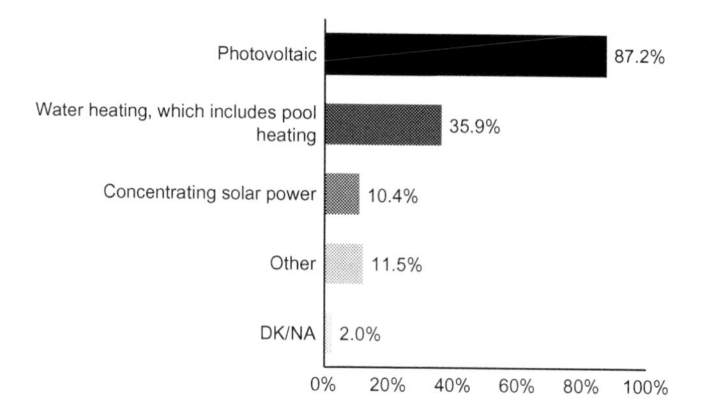

According to the 2012 Solar Foundation Solar Jobs Census, installation firms, which are the most labor-intensive of all solar companies, make up the largest segment of the US solar industry at 48% of all employment in 8813 establishments. Manufacturers follow installation at 25% (1262 firms), which are increasingly struggling with falling module prices. Sales and distribution firms stand at 13.4% of employment at 3050 establishments; all other, including R&D at 6.4% and 1454 firms.

Subsector	2011 Jobs	2012 Jobs	2011–2012 Growth Rate	2013 Projected Employment	2012–2013 Expected Growth Rate
Installation	48,656	57,177	17.50%	68,931	21%
Manufacturing	37,941	29,742	−21.60%	32,313	9%
Sales and Distribution	13,000	16,005	23.10%	19,549	22%
Project Development	–	7,988	–	9098	14%
Other	5548	8105	46.10%	9551	18%
Total	105,145	119,016	13.20%	139,442	17%

[4]Due to many firms working with more than one technology, the sum is >100%.

From 2011 to 2012, most of the growth (86%) came from new hires, as opposed to shifting existing workers from other activities to solar ones. These new hires were spread throughout various categories, but most were not lower skill or lower experience jobs. In fact, over half required previous experience related to the position, 41% required a bachelor degree or more, and 18% required an associate degree or certificate (but not a bachelor degree). Interestingly, only 2% required union membership.

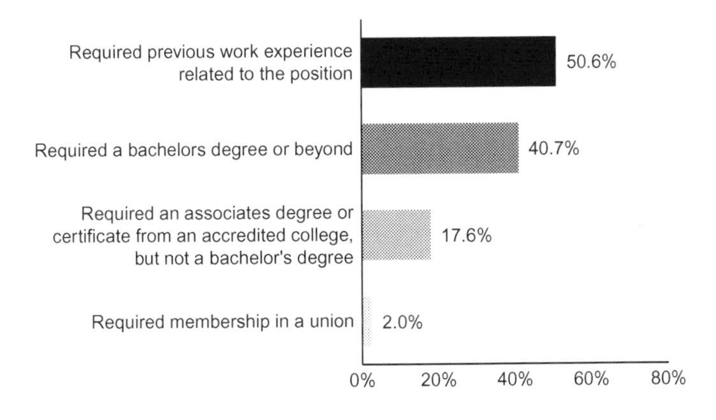

The types of new jobs were most frequently added in the production/technician function group (41%), followed by management and professional jobs (23%). Sales and administrative functions came in around 15% each.

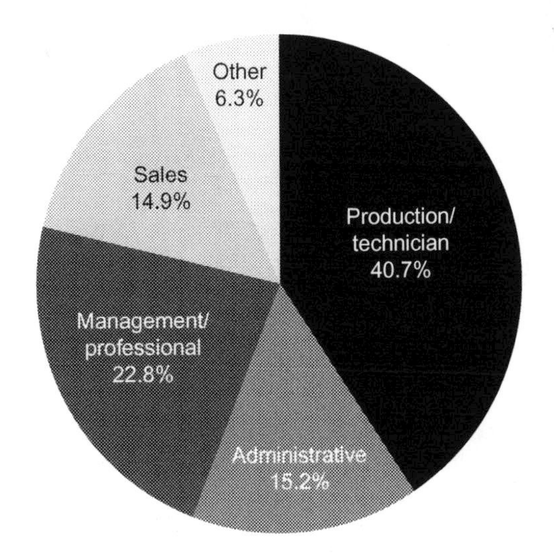

One particularly interesting finding is that employers are not terribly interested in most traditional job search methods. Only 16% use online job postings, and only about 5% recruit directly from schools and colleges. Fifty-four percent, on the other hand, prefer word of mouth or referrals. Even as it matures, it is clear that the solar industry does not want to take unnecessary risks on its workers.

About one-third of employers in 2012 had no difficulty at all finding qualified workers, and another 53% said they had only some difficulty (9.6% reported great difficulty). This is surprising given the rapid growth of the industry, however, the explosion of US-based training programs—particularly in solar installation—together with high unemployment from the Great Recession explain the relative ease of hiring qualified workers. This is a trend that is unlikely to continue as demand for solar increases, stimulus-funded training programs dwindle, and the general economy continues to add jobs.

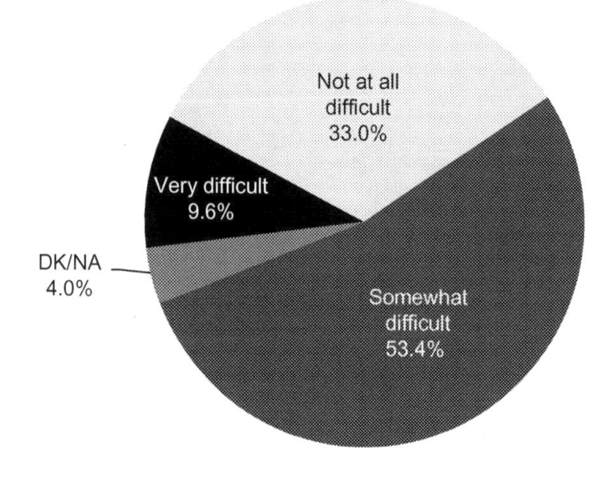

Installation

The installation sector makes up the biggest segment of the US solar industry. As of 2012, installation firms employ 57,177 solar workers at 8813 establishments. The majority of these, over 6500, are quite small, employing only 2 or 3 solar workers and are engaged in other work in addition to installation.

Installation employment has grown dramatically over the last several years. In 2010, the first year of reliable and comparable estimates, installation firms employed just over 43,000 solar workers. This number is expected to swell to nearly 70,000 by the end of 2013, a staggering 57% growth rate over 3 years.

Installers added the most new solar workers of any solar sector in 2012, more than offsetting declines in US solar manufacturing. While most installers are small, larger firms are growing faster (in both real numbers and percentage growth), further suggesting consolidation and maturation of the subsector.

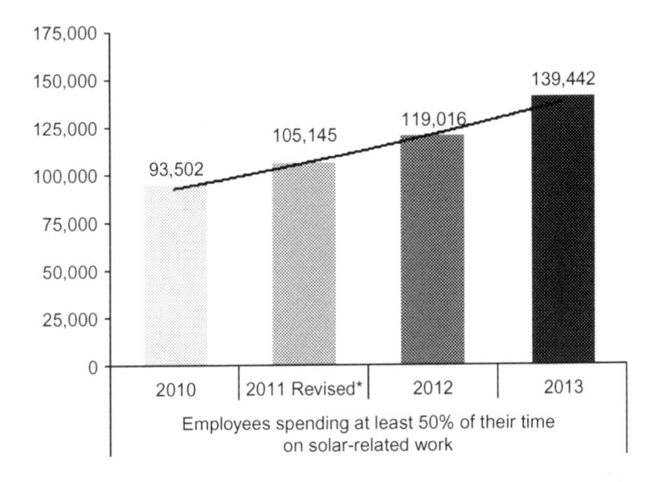

Firms are also bullish on growth with a plurality of 44% expecting to grow in 2013.

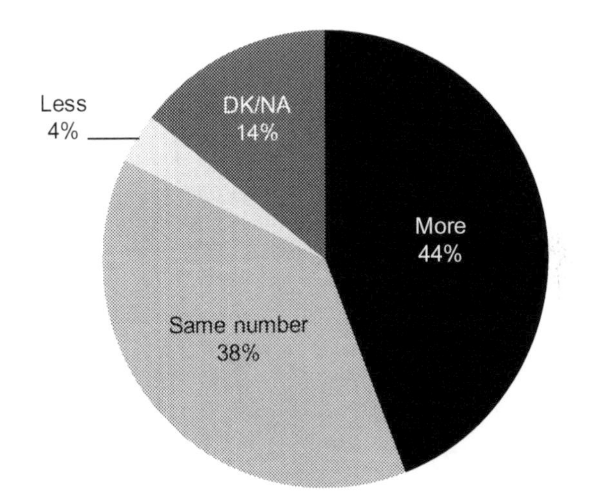

In addition to these growth trends, installation firms are more likely to be specialists now than they were even 3 years ago. A majority of firms surveyed for The Solar Foundation's 2012 Census reported that they receive all or most of their revenue from solar projects, rather than solar only accounting for a portion of total revenue. 2012 was in fact the first year that firms became so specialized, as previously, fewer than 50% of installers earned all of their revenue from solar. This consolidation is a hallmark of maturing industries.

As of September 2012, 51% of installation firms receive 100% of their revenues from solar installations, up significantly from 39% the previous year. Eighty percent earned a majority of their revenues from solar, up from fewer than 6 in 10 in 2011.

Photovoltaics dominate the landscape with only a handful of firms working with CSP. However, this is as much a function of the size of systems that firms work on with the overwhelming majority of installers focused on smaller systems. In fact, approximately 7750 firms in the United States work on residential systems while only 325 perform installations of utility-scale power systems. As a result, 91% report working on PV systems and just under 5% work on CSP.

The solar installation labor market trends show increasing specialization. Installers in the United States are more likely to either hire in-house specialists or contract out portions of their solar work than in years past. This includes administrative positions, on-site installation, electrical, or construction work, and other positions related to the installation process. This is in stark contrast to even 5 years ago when firms tended to rely more on cross-functional employees to conduct these multiple and varied tasks.

The resulting specialization appears to be producing greater efficiency in the installation process with more specific work tasks and fewer people who are expected to conduct multiple installation activities (such as assessment, customer service, and rooftop installation).

Several data points confirm this analysis. In the 2012 Census, solar installers were asked detailed questions about their work activities, in addition to their preferences for employees. The data show that from 2011 to 2012, the average solar system size increased, but the average amount of labor hours for a system installation did not.

Installers across the entire spectrum of applications and sizes reported that the average time required to complete an installation increased by just over 2 h on average or <1% of the total time to complete a system. At the same time, the average installed system grew dramatically (residential systems alone grew in size at twice the rate of the labor rate).

The second key data point focuses on the productivity of solar workers in the installation sector. In 2010, the installation workforce of 43,934 installed 956 MW of solar power, 887 MW of which were PV systems. This averages to 46 workers per megawatt of installed energy. Especially because 2010 was a year with fewer distributed generation installations, this number is very high, showing an inefficient solar workforce.

The 2010 figures contrast starkly with 2011 figures. In that year, 48,656 workers installed 1855 MW of solar power.[5] In other words, 11% more workers were required to install 109% more solar energy! The 2011 figures show an impressive drop to 26.2 workers per megawatt.

In the following year, installers ramped up their hiring at an even faster pace, adding about 8500 new installers at an 18% clip. At the same time, solar installs nearly doubled again with 3343 MW of solar installed. The resulting labor market intensity is a much smaller 17.2 workers per megawatt.

[5] http://www.seia.org/news/new-report-finds-us-solar-energy-installations-soared-109-2011-1855-megawatts.

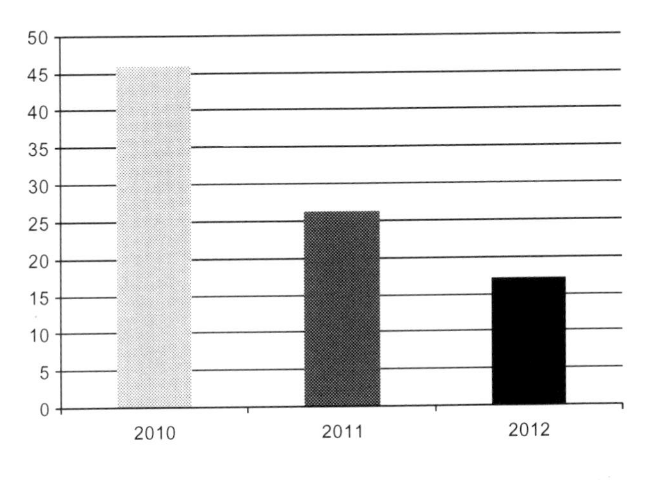

Several factors are important to consider. First, nearly all of the installed capacity has been PV, yet several large CSP (and utility-scale PV) plants have been under construction. As a result, it is not a perfect analysis because clearly a portion of the 51,000 US installation workers are working on systems that are not yet online.

At the same time, labor efficiency is directly correlated to system size. According to SEIA, "of the 878MW of PV installed in the US [in 2010], 45% were for non-residential installations (including commercial, non-profit, and public projects) and the remaining portion was divided equally between utility and residential solar installations."[6]

In SEIA's 2011 report with GTM Research, the utility-scale/non-residential growth story continues: "The utility market grew 185 percent in 2011 to reach 758 megawatts, accounting for 41 percent of all installations in 2011. There are over 9,000 megawatts of projects with signed utility PPAs in the pipeline, and over 3,000 megawatts currently in construction."[7]

The year 2012 began a shift in the solar industry, where residential growth outpaced nonresidential solar growth. This is led in large part by third-party ownership, which is responsible for nearly all of this growth (host-owned systems are essentially flat). As these systems gain popularity, we must continue to watch productivity curves, as the total labor intensity for small, distributed generation projects is significantly higher than large, commercial, industrial, or utility projects, by simple matter of efficiency of scale.

Manufacturing

US solar manufacturing is in a tailspin. With the exception of a few larger, niche firms, the substantial oversupply and resulting precipitous declines in module prices have led to sharp declines in the industry.

From 2011 to 2012, US solar manufacturers shed about one-fifth of their workforce. Several high profile bankruptcies, including the well-publicized Solyndra and

[6] http://www.brightstarsolar.net/2011/03/us-solar-installed-capacity-sees-fastest-growth-in-2010/.
[7] GTM Research/SEIA, 2011 US Solar Market Insight.

Evergreen Solar cases, were only the tip of the iceberg. While these firms could easily be chalked up to poor management or lack of demand for innovative, next generation products, finish product manufacturers throughout the country faced declines.

As of 2012, 1262 solar manufacturing establishments in the United States employed 29,742 solar workers. Forty-one percent of firms produce finished products only, and these firms suffered the greatest declines. The 31% that solely manufacture components also saw some declines but their struggles were much less dramatic than the finish product manufacturers.

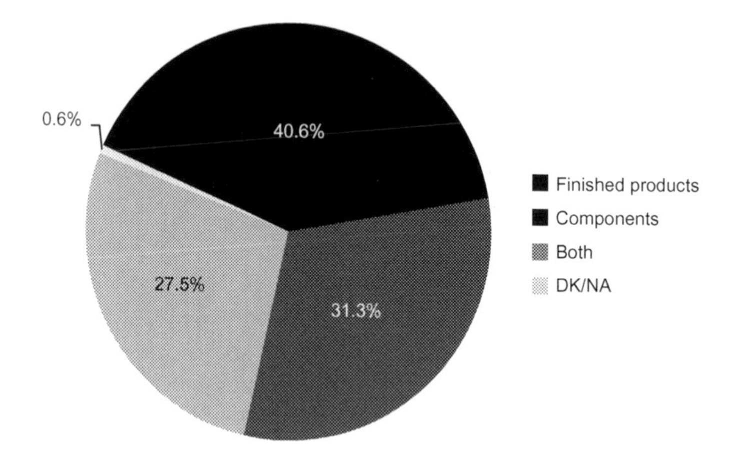

Seventy-five percent of manufacturers provide goods for the PV market, which has grown dramatically. In fact, the figure was up significantly from 2011, when only 59% produced for PV. The increase in PV manufacturing (as opposed to CSP or solar water heating) is likely due to two factors. The first is that PV has grown more quickly than the other technologies and the second is that US manufacturers provide many supplies and subcomponents to foreign PV manufacturing firms, so global growth in PV manufacturing benefits these manufacturers. If the latter is the case, then the future looks murky even for component manufacturers, as the oversupply of PV in the market will ultimately lead to diminished demand for components.

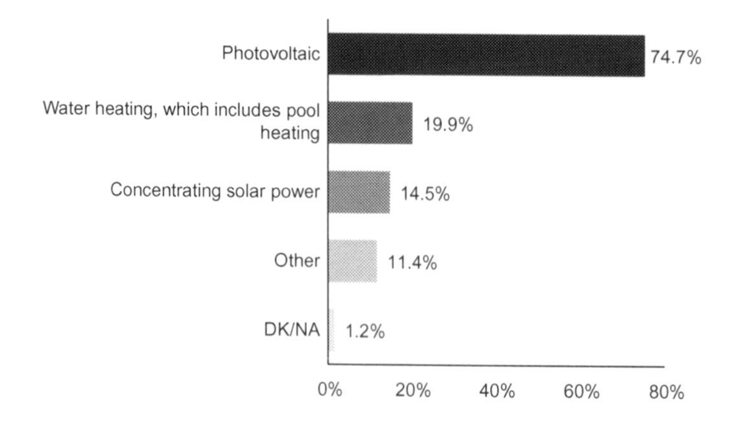

As seen in the figure below, US component manufacturers are most heavily focused on inverters, racking, and modules. This may reflect an overall trend for US manufacturing of continued strength in precision manufacturing.

Manufacturing has significantly lower labor market intensity than installation in the solar sector. Investments pour into efficiency in the manufacturing process to reduce labor costs. Automation, robotics, and other laborsaving mechanisms have driven down the hours associated with production. As a result, jobs in direct production do not drop as dramatically as the decline in sales.

Despite this, there are numerous nonproduction, back office jobs that have been devastated by the downturn in US solar manufacturing. These jobs include support, logistics, administrative, sales, and management.

From a production and output standpoint, the US manufacturing sector is in clear decline. In 2010, US solar manufacturers produced 1273 MW of PV. As noted earlier, the manufacturing cluster has been in flux for several years, and there is overlap in technology applications within firms, making the labor market intensity question a bit more challenging. Based on The Solar Jobs Census series, it is clear that PV is responsible for a greater share of the manufacturing sector, representing approximately 59% in 2010, 54% in 2011, and 64% in 2012.

For the purposes of the labor intensity analysis, these figures translate to approximately 23,600 PV manufacturing workers in 2010, 20,500 workers in 2011, and 19,000 in 2012. The resulting labor market intensities are therefore:

2010—18.5 workers per megawatt and
2011—16.8 workers per megawatt.

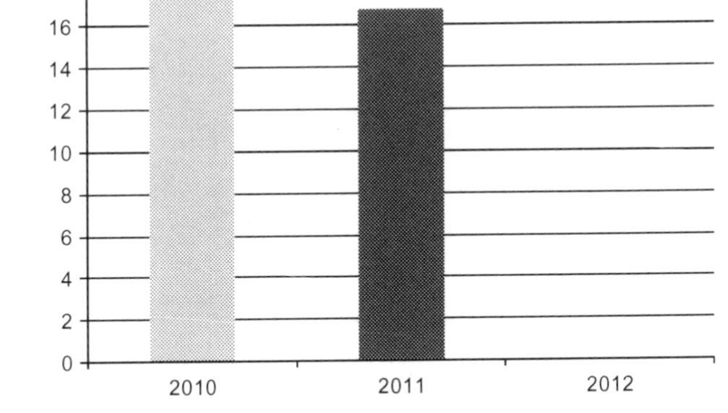

Sales and Distribution

Sales and distribution firms are an important bellwether for the solar industry as these firms are engaged with both domestic and foreign imports and exports. As of the third quarter of 2012, there are 3050 solar sales and distribution establishments in the United States employing 16,005 solar workers.

The majority of sales and distribution companies focus on PV systems and solar water heaters; however, a trend is emerging with more firms selling or distributing PV and significantly fewer offer solar water heating products. This is important to consider in the future, as solar water heating continues to lag PV in uptake.

Slightly more than 1500 of the firms in sales and distribution are strictly solar businesses. The others also work with other electronic and heating and cooling systems. Over time, as the industry matures, more and more firms are devoting their activities exclusively to solar products.

Project Developers

Solar project developers refer to the class of companies that oversee the construction of larger scale installations. These firms, often large construction or energy companies, typically manage the entire process from site preparation, obtaining financing, managing legal issues, and ultimately building the plants. Some developers also manage the systems, though the operation and management of utility-scale solar requires minimal labor inputs.

There are just over 400 companies identified as solar project developers in the United States employing nearly 8000 solar workers. Of these, nearly 95% work with photovoltaics, while about 10% work on CSP.

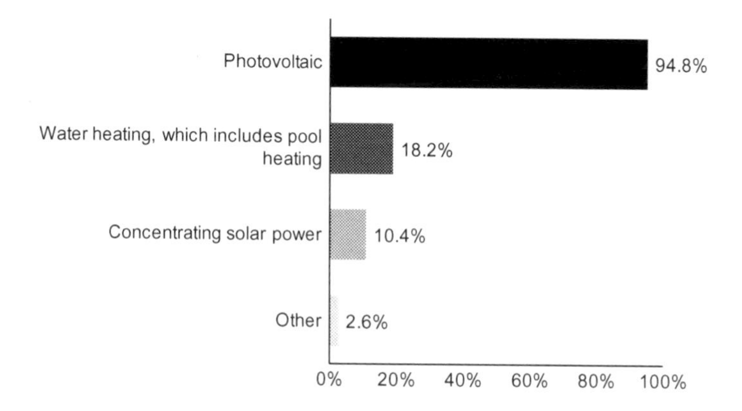

Other

The solar industry has many other types of firms that do not classify neatly into manufacturing, installation, sales, or project development. These firms include research and development organizations, financiers, law firms, and others. Employers in this category contribute over 8000 solar workers to the labor force of the industry.

Supportive firms are not as frequently working exclusively with solar. Nearly one-third earns <50% of their revenue from solar.

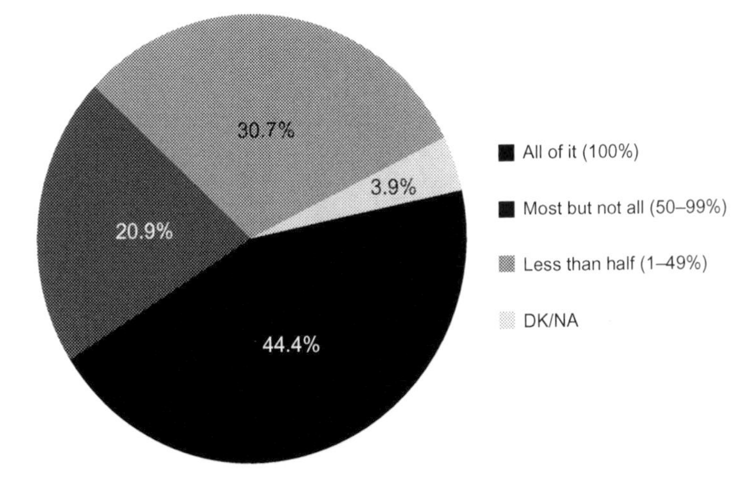

In addition, the overwhelming majority (87%) work with PV products and services rather than water heating or CSP.

Germany

The German labor market is quite different from its global competitors. It is much more efficient, as the industry in Germany has overall lower soft costs and generally larger systems. As noted in Chapter 5, 42% of Germany's 2012 PV installations were >1 MW,[8] so just by nature of efficiencies of scale, we would expect lower labor intensity.

While Germany has long been an efficient labor market, Germany Trade and Invest reports that its productivity has increased dramatically since 2005.[9] This increased efficiency translates to a yearly average decrease to labor unit costs of 0.3% for the period 2005–2010.[10] The report credits "highly flexible working practices, such as fixed term contracts, shift systems, and 24/7 operating permits" for these decreases, particularly in the face of rising costs elsewhere.[11]

As noted repeatedly in this text, the German installation sector is significantly larger and has grown at a much faster rate than the rest of the world. It has outpaced the United States by about 2.5 times from 2005 to 2011, and has about 3.6 times

[8] http://www.gtai.de/GTAI/Navigation/EN/Invest/Industries/Energy-environmental-technologies/solar-industry.html.
[9] Germany Trade and Invest, Industry Overview: The Photovoltaic Market in Germany, Issue 2013/2014.
[10] *id.* at p. 11.
[11] *id.*

more cumulative solar PV installed.[12] This alone is likely to represent about half of the differences in soft costs between the United States and Germany.[13]

Germany is often thought of as a production state, however, recent declines in manufacturing based on pressure from China and collapsing subsidies have devastated the market. In 2011 and 2012, at least 12 German cell manufacturers filed for bankruptcy. Official statistics state that at the end of 2012, only 6000 workers were employed in the sector. This is a dramatic decline as it represents between 25% and 30% of the industry.

The story is the same all across the globe. Even in the face of dramatic demand-side increases, the supply of panels remains too high. Whether based on overzealous production, global competition, market manipulation, or other factors, the glut of product has prices in a sustained free fall.

Employment data are less granular in detail for Germany than for the United States. At the time of print, it is estimated that the PV industry in Germany employs 100,000 workers. Given that the overwhelming majority of the 120,000 workers in the United States are engaged in PV work, and that Germany installed nearly twice the PV that the United States did reflects just how much further along Germany is in terms of labor efficiencies than the United States. Of course, this analysis says nothing of component, materials, or finished product manufacturing, nor does it address logistics, sales, or other categories. However, given all the data, it is clear that Germany is at least 30% more efficient than the United States on a jobs-per-megawatt basis.

As installations peak and global supply for product create headwinds, Germany is poised to compete in the next generation of PV, the thin film market. This optimism stems from several data points. First, the weakest competition has fallen, and remaining companies are stronger. Second, thin film offers expanded opportunities for installations, which benefit German building stock. Finally, Germany has tremendous human capital assets.

The human capital assets stem from the nation's long history as a production giant, as well as its more recent strength in solar energy. Germany has a highly skilled and specialized market, including numerous engineers and technicians. Its 312 PhD graduates per million inhabitants rank second among OECD nations.[14] There are now 300 renewable energy courses at universities and connectivity with other high-tech sectors.[15]

Germany also has an intriguing dual education system, in which students get training and experience. This has been noted as a critical combination in employer surveys in the United States and abroad. This system offers high quality skill development and provides employers with easy access to pre-screened talent.

The investments in human capital and the differences in labor wages draw sharp distinctions with the United States. According to analysis by the Lawrence Berkeley National Laboratories, some US jobs pay significantly higher wages in the sector,

[12] Seel et al., Why are residential PV prices in Germany so much lower than in the United States? A scoping analysis. Available from http://emp.lbl.gov/sites/all/files/german-us-pv-price-ppt.pdf, p. 6.
[13] *id.*
[14] Germany Trade and Invest, Industry Overview: The Photovoltaic Market in Germany, Issue 2013/2014.
[15] *id.*

including those in skilled technical fields, such as electrician. These occupations have scattered training and there is a shortage of such skilled labor in the United States, as opposed to Germany which does a much better job at aligning supply with demand. However, the same equilibrium appears to have increased wages in some nontechnical categories, such as administrative tasks, vis-à-vis the United States.

Overall, the German labor market's efficiency and diversity of technology and value chain strength should weather the short-term volatility in the industry. It is likely that after a few years of flat to declining year over year employment, the job market will rebound and provide long-term, stable, employment.

China

There are no reliable estimates on the size and scope of Chinese solar industry employment. However, there are indicators that allow for some useful analysis of the industry.

First, China is heavily invested in panel production over installation. With the help of state run banks and massive public investments in infrastructure and demand, China has risen to take 80% of European solar market.[16]

Chinese solar production increases have helped to fuel the global solar module output from only 100MW in 2000 to approximately 50,000MW in 2012, a 500-fold increase in only 12 years! According to Steven Lacy, "In four years, the solar manufacturing sector shifted from being led by a geographically dispersed number of companies to one dominated by Chinese companies. In 2006, there were two companies from China in the list of top ten cell producers. In 2010, there were six, according to Bloomberg New Energy Finance (BNEF). There are currently only two non-Asian manufacturers in the top ten, and those companies – First Solar and Q-Cells – have shifted a lot of their production to Asia."[17]

As noted in Chapters 5 and 9, this dominance is nearly certain to continue, but only because China's decline from restructuring and consolidation will be less dramatic than in Europe and the United States. However, the current pace of growth, price declines, and government support are not sustainable, and a significant correction is on the horizon, unless China dramatically increases its domestic demand for solar panels.

Suntech, one of the world's largest PV manufacturers, has filed for bankruptcy, potentially wiping out 10,000 Chinese jobs in the process. LDK Solar is also in serious financial trouble, particularly jeopardizing the 40% or so of their 25,000 workers who work in wafer production. The Chinese government, meanwhile, appears ready to allow reduced production to bolster prices and force additional consolidation. This is widely viewed as healthy for the solar industry in the long term, despite the likelihood of increases—or at least much slower decreases—in the price of solar products.

[16] http://www.europeanvoice.com/article/2013/june/breaking-news-eu-imposes-tariffs-on-chinese-solar/77418.aspx.

[17] http://www.guardian.co.uk/environment/2011/sep/12/how-china-dominates-solar-power.

The Chinese government could certainly learn a lesson from the United States in its recent labor market transformation. By promoting installation over production, demand increases, bolstering the entire industry. As seen in the United States over the last several years, the installation sector is much more labor intensive, so from an employment perspective, it has much more potential to create jobs.

8 Global Markets

After a decade of strong growth, the global solar market—from a production and profit standpoint—has stalled. The financial crisis in the United States sent domestic markets reeling, and reduced government tax receipts have led many states to question their often generous incentives for renewable energy. As states have been reluctant to expand their own incentive programs in the United States, the impact of the federal government's aggressive stimulus plan, which included numerous market-driving policies for the solar industry, is subsiding.

While the storm clouds over the US economy appear to be lifting, Europe—the leading region for solar-installed capacity—remains in turmoil. The austerity measures put into place to right the economic ship seem to be only exacerbating EU financial woes. These same austerity measures have drastically limited solar market expansion in the near term.

Compounding the solar industry's challenges, an aging grid in developed markets, transmission project delays, and a highly charged global policy debate worsen the outlook over the coming years. These issues, together with a substantial oversupply of photovoltaics, have led many firms—particularly in the manufacturing sector—to fold.

Despite these seemingly many pitfalls, there are some global bright spots. Competition is driving down costs and increasing efficiency. Emerging markets are developing thoughtful programs and policies with more proven track records. New, well-financed market entrants, such as the United Arab Emirates and Saudi Arabia, are investing heavily in the industry.

A review of the expected trends facing the energy sector is critical to understand the future of solar markets. This chapter begins with a focus on global energy markets and then reviews the major themes relative to solar power in several exemplary countries.

Global Energy Demand

As the world's population continues to soar, energy demand is growing at a dramatic pace. Fueled in large part by Asia, and specifically China and India, global demand is expected to increase 53% between 2008 and 2035. Members of the OECD, which tend to be more established economies, are expected to grow by only about 0.6% per year, while non-OECD states at a rate nearly four times higher.[1]

[1] http://www.eia.gov/forecasts/ieo/world.cfm.

Solar Energy Markets. DOI: http://dx.doi.org/10.1016/B978-0-12-397174-6.00008-8

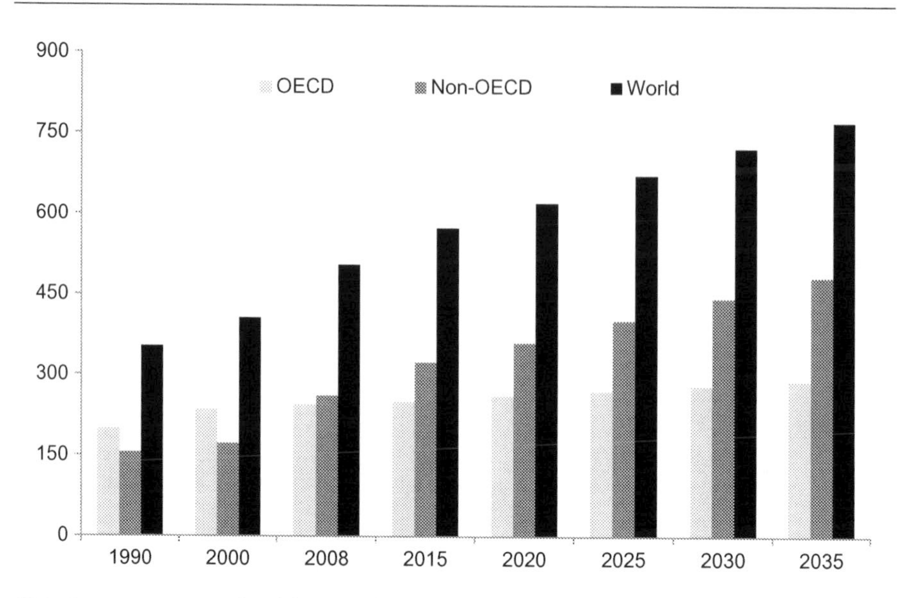

Global energy consumption, Energy Information Administration (EIA) estimates

Recent history illustrates this eastward shift. In 2009, as the United States began to emerge from its recession, its energy consumption declined by 5.3%. In that same year, China's energy consumption is estimated to have eclipsed the United States for the first time. The US Energy Information Administration models predict that by 2035, China's energy consumption will be 68% higher than the United States.[2]

While growth in renewables is expected to be significant, it is roughly in line with other technologies. Liquid fuels like oil and nuclear are expected to grow more slowly, while coal and natural gas will continue a solid gain.

EIA categorizes energy consumption by four areas: transportation (mostly dominated by liquid fuels), residential, commercial, and industrial. These categories allow for more refined and accurate projections and are also very important when considering the types of solar energy systems that will have the most important impacts across the globe, as well as highlighting the best target markets for each.

Residential

Residential energy consumption refers to household use, which is comprised mostly of heating fuel and home electrical use. As would be expected, the developed, OECD member states will see much slower population growth, and therefore relatively small annual increases of energy consumption.[3] Non-OECD states, particularly in Brazil, India, and China, are expected to grow much more rapidly at a rate sevenfold higher.

However, since the EIA estimates were produced, advances in efficiency, slower economic growth, and relatively high-energy prices have led to a surprising result. Even as the United States emerges from recession, its energy consumption is not

[2] *id.*
[3] *id.*

following suit. As a result, while energy consumption is expected to increase over time, the comparison to Asia is even more dramatic. This is reflected in the EIA 2013 outlook that predicts that US energy consumption over the next 30 years will remain flat.[4]

These trends have important implications for the solar industry, including:

- In the United States and Europe, larger homes that consume higher amounts of energy dominate, but use will not grow dramatically.
- Residential solar in the United States and EU will be more heavily focused on driving down electrical costs rather than meeting demand.
- Grid-connected residences in Asia will require new sources of energy and as demand grows, solar electric systems will compete with the cost of building new systems and transmission rather than merely adding more fuel.
- Non-grid connected residences in Africa, the Middle East, and Asia will seek new sources of energy as household incomes rise and the use of wood and other biofuels diminishes.

Commercial

Commercial energy use includes the consumption of the service sector. It is highly connected to population growth (as services tend to population serving). Commercial energy use is driven by electrical, heating, and cooling of buildings and other structures, though traffic lights, water, and sewer systems are also included in this category.[5]

The EIA expects commercial energy use to expand by an annual 1.5% globally. Energy use will grow more quickly in developing nations, as OECD states expect much slower population growth and faster adoption of energy efficiency measures.[6]

Industrial

Industrial energy use is predominantly used for production, as well as for lighting and other business uses in the manufacturing industries. Industrial production makes up just over half of all global energy use and is expected to grow by a similar 1.5% globally each year through 2035.[7]

Again according to EIA, "not only because of faster anticipated economic expansion but also because of the composition of industrial sector production. OECD economies generally have more energy-efficient industrial operations than non-OECD countries, as well as a mix of industrial output that is more heavily weighted toward non-energy-intensive industry sectors. As a result, the ratio of industrial energy consumption to total GDP tends to be higher in non-OECD economies than in OECD economies. On average, industrial energy intensity (the consumption of energy consumed in the industrial sector per dollar of economic output) in non-OECD countries is double that in OECD countries."[8]

[4] http://consortiumnews.com/2013/07/09/us-energy-renaissance-shifts-geopolitics/.
[5] *id.*
[6] *id.*
[7] *id.*
[8] *id.*

Global Renewable Energy Outlook

Several newly developed energy future scenarios have been developed with somewhat different results. The future is difficult to predict across all energy markets and significantly more so for renewable energy ones. Analysis requires assumptions for GDP, the policy landscape, continued environmental awareness, and the investment landscape.

In a recent report by BNEF, a "new normal" scenario is outlined with world economic growth at pre-recession levels, stronger policy coordination among major emitters, sustained investments, continued technological innovation, and strong demand for fossil fuels.[9] In this scenario, BNEF estimates that by 2030, renewable energy will make up nearly half (48%) of total power generation (up from 28% in 2012).[10] Solar is expected to make up between 14% and 17% of global energy production under the various future scenarios.

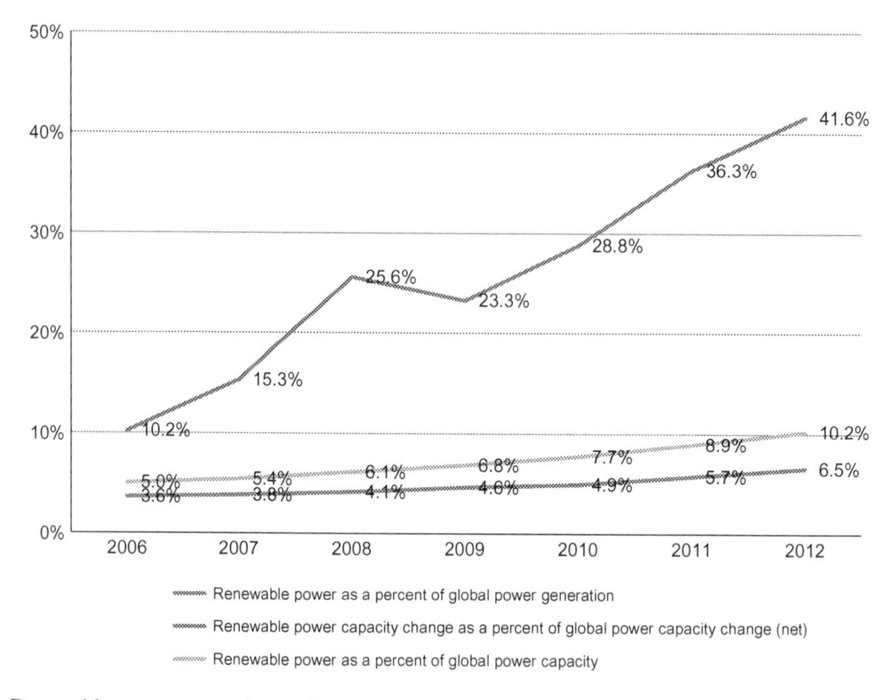

Renewable power generation and capacity as a proportion of global power, 2006–12
Source: http://www.unep.org/pdf/GTR-UNEP-FS-BNEF2.pdf

This figure suggests a continued rapid expansion of solar energy. In fact, BNEF estimates that solar PV will account for 30% of all new power additions through 2030. Solar is expected to become the leading renewable source by share of market, rising to account for between 27% and 32% of all renewables by 2030.[11]

[9] Bloomberg New Energy Finance, Global Renewable Energy Market Outlook 2013, Fact Pack, April 26, 2013.
[10] *id.* at p. 3.
[11] *id.*

This rise is quite dramatic and in line with estimates that suggest rising fuel prices. BNEF also estimates that 73% of total investment in power generation will be directed to renewables, to the tune of nearly $8 trillion. About $5.5 trillion of this will be invested in solar and wind with a maximum of $3.2 trillion under the most aggressive scenario for solar alone.[12]

These long-range scenarios are fascinating, but shorter term projections have greater accuracy. Renewable energy production has grown by a 5.8% rate globally in 2010 and 2011, leading the EIA to project a growth rate for 2011–2017 that is 60% higher than 2005–2011. In its scenarios, it expects the number of countries that produce solar PV to double between 2011 and 2017.[13]

Overall energy demand is clearly important for the proliferation of solar technologies. Increasing global demand, concern over climate change, increasing competitiveness of solar economics, and strong public support have led to a massive increase of installed solar electric capacity across the globe.

EIA projects that by 2017, solar will contribute 4.9% to renewable generation. This is based on an expectation that solar PV will grow by 35 TWh per year—an increase of 27.4%—from 70 to 230 GW over the period. This growth will be led by China (32 GW), the United States, (21 GW), Germany (20 GW), Japan (20 GW), and Italy (11 GW).[14]

CSP is expected to explode over the period from 2 to 11 GW by 2017.[15] Despite this 550% growth scenario, expectations were higher and the small base tempers the excitement. Clearly, PV growth has eroded some of these gains. Nonetheless, the United States is expected to lead all nations in CSP with 4 GW added by 2017. At the same time, solar thermal heating is expected to grow by 150% over the period.[16]

These global scenarios illustrate a strong and continued demand for solar energy. A country-level analysis is also critical for determining the best markets for solar in the future.

According to the SEIA, US annual PV installations grew to an impressive 3313 MW in 2012, an increase of 76%. The new capacity represents 11% of global installations, its highest share in 15 years, and increases the US installed capacity figure by more than 100% from 2011 at 7221 MW.[17]

The residential market is booming in the United States with over 83,000 installations in the sector. This is largely driven by cost; the blended average sales price for PV modules for Q4 2012 was $0.68/w, a staggering 41% below the Q4 2011 price of $1.15/W.[18] This pricing led to further reduction to the cost of PV systems and an expectation of 30% growth in 2013.[19]

The 2012 growth numbers are significant, especially given that over 39% of the capacity was installed in Q4. The market was led by California, the historical leader, as well as New Jersey, Arizona, North Carolina, and Massachusetts.

[12] *id.*

[13] Energy Information Administration, Medium Term Renewable Energy Market Report 2012, p. 11. Available from http://www.iea.org/textbase/npsum/mtrenew2012Sum.pdf.

[14] *id.* at p. 14.

[15] *id.*

[16] *id.*

[17] U.S. Solar Market Insight Report: 2012 Year in Review, Solar Energy Industries Association and GTM Research, Executive Summary.

[18] *id.*

[19] *id.*

In the United States, the residential market continues to grow steadily with installers adding 488 MW in 2012, growth of 62% from 2011. The fastest growing residential markets are California, Hawaii, Arizona, and Massachusetts. Only Pennsylvania, which suffered from renewed policy uncertainty, shrank in 2012.[20]

The nonresidential market in the United States grew 26% in 2012 adding 1043 MW of new capacity. The market leaders, California and New Jersey, both started 2012 strong but declined after midyear. This was offset by emerging growth in Massachusetts, Hawaii, Maryland, and New York.[21]

While these figures are impressive, utility growth exploded, more than doubling from 2011 at 1782 MW (more than half of the year's installed capacity growth). The rise, however, has meant that many states have hit their renewable targets, so growth is expected to slow in most parts of the country.[22]

GTM Research projects 2013 PV capacity additions to grow by 29%, again outpacing the global installation rate but down from 2012 impressive rate. The resulting capacity additions total 4.3 GW of PV. This is mostly lower due to slowing utility installations, and also likely to track previous years, with strong Q1 and Q4 installations. The projections seem on track, with 723 MW reported to be installed in Q1 of 2013, representing 33% growth over Q1 2012.[23]

Ernst & Young and BNEF expect that public opinion, as demonstrated in the reelection of President Obama and recent polling, will drive the United States to continue its growth in the medium term for solar energy. Americans grow wearier and wearier of reliance on foreign oil and its inevitable entanglements, and are seeking support for domestic sectors that are likely to grow jobs.[24]

The United States leads the Ernst & Young solar attractiveness index, maintaining its position in 2011 and 2012. India remains at #2, a surprising ranking for a country with an old and insufficient grid, blackouts, policy uncertainty, and other issues. However, with one-third of India's population off the grid, and a significant energy deficit in the country, there is potential for significant growth in distributed generation.[25]

China ranks third in the indices, however, the shift towards the nation is fully underway and if current trends continue, it will easily lead the world in the near term. While solar has not taken hold as quickly as wind power, for example, but changes in policies and new investment appear to be rapidly changing the focus of the country to installed solar capacity. Much of the shift in focus is to further stimulate the saturated PV production plants.[26]

Europe continues to fade, as it reaches near saturation in most of its markets. Germany, a perennial top nation in the index, sits at 4, while Spain has fallen to 9.[27]

Several new markets have emerged on the scene, however, predominantly in the Middle East. Specifically, the UAE and Saudi Arabia are marching ahead, both making the index for the first time in 2012 at #12 and #14, respectively.[28]

[20] *id.*

[21] *id.*

[22] *id.*

[23] http://www.seia.org/research-resources/solar-industry-data.

[24] Ernst & Young, Renewable Energy Country Attractiveness Indices, November 2012, at p. 39.

[25] *id.* at p. 29.

[26] *id.* at p. 26.

[27] *id.*

[28] *id.* at pp. 34–35.

Of course, these statistics and indices are useful to analyze the markets for installing solar capacity, but do not address the market for production. This is because, quite frankly, over the course of time it has taken to write this text, the section highlighting areas of opportunity in production has shrunk to almost nothing.

Global public investments in renewable energy companies dropped a precipitous 60% in 2012 continuing a free fall from the peak in 2007. Currently, public investment in the sector sits at only one-fifth of the 2007 high. Solar investment fared slightly better than other renewables and claimed the top spot among all technologies, however, this is only because others declined faster. In 2012, solar investment dropped 50% to $2.3 billion globally.[29]

In 2012, global production of PV reached 60GW for 30GW of demand.[30] A PhD in economics is not required to recognize the resulting impact on prices with continued slides and lost profitability. Crystalline silicon module spot prices declined from about $1/watt to an average of $0.80/watt and even as low as $0.60/watt on some larger deals.[31]

The resulting slide in prices decimated profits along the value chain with nearly all producers losing money. Many public companies reported staggering losses, while 2012 saw the demise of Q-cells, the German manufacturer that at one time was the largest producer in the world, Centrotherm, a German equipment maker, and Konarka in the United States. Traditional PV was not alone in this mess; Abound Solar, once a darling of the US Department of Energy (and recipient of guarantees similar to those received by the media sensation Solyndra) and producer of thin-film modules, also declared bankruptcy.[32]

It is no surprise that clean energy stocks suffered in 2012, but it is perhaps surprising, given the pain in the supply chain, at how little they were down in 2012—5.9% according to The Wilderhill New Energy Global Index (NEX). The 2012 NEX valuation of 120.02 is down 78% since 2007 and is clearly underperforming against the wider market. Despite these bad conditions, solar remained the largest issuer of shares through Initial Purchase Offerings (IPOs) and still managed to attract significant investment over the period.[33] The stock valuation was clearly also supported by strength among large developers who are turning in record profits.

The year 2012 ended with a bang, a trend that seems to have continued into 2013. The NEX added 15% from December 1, 2012, to January 31, 2013, with the Bloomberg solar index adding 5%.[34] However, the rebound may be a red-herring and not all commentators agree on the sustainability of the rebound.

The solar industry is clearly offering investors and market analysts mixed messages. As Michael Liebreich, the chief executive of BNEF put it, "for every equipment company operating at thin or negative margins, there is an installer who is getting a good deal… Rumors of the death of clean energy have been greatly exaggerated."[35] Ultimately, however, solar will not see market competitiveness *industry wide* until equilibrium is found between installation and production.

[29] *id.* at p. 56.

[30] http://www.unep.org/pdf/GTR-UNEP-FS-BNEF2.pdf at pp. 57–58.

[31] *id.* at p. 58.

[32] *id.*

[33] *id.* at p. 59.

[34] *id.* at p. 60.

[35] http://www.bloomberg.com/news/2012-01-12/clean-energy-investment-rises-to-a-record-260-billion-on-solar.html.

9 The Economics of Solar Power

Energy markets are driven by economics. Throughout history, most attempts to internalize externalities have been fruitless. In the case of energy, it remains highly unlikely that an agreement for carbon pricing could be achieved—either globally or with the two largest emitters, the United States and China. Given this reality, the future of solar will only be as bright as its competitive stance vis-à-vis competing energy technologies.

Each of the previous chapters has addressed the critical components to the solar industry's competitiveness. This chapter seeks to untangle the complicated—and often misleading—discussion on price parity, the point at which solar energy becomes truly cost competitive with fossil electric power production. At the outset, however, it is important to note that total cost is not the only important factor to consider. A safe and reliable electricity market requires diversity of fuel sources, peak-load management, and low maintenance, fixed priced production, all of which solar power offers in spades.

The economics of solar power are sufficiently complicated and difficult to understand, however, the amount of confusion—and outright misinformation—about the price of solar electricity is alarming. The general public, policy makers, and even some academics seem to be operating in an alternate reality of decades past, where solar electricity could not possibly compete with traditional energy sources.

In reality, price competitiveness is much closer than many think and has already arrived in many markets, especially those with strong subsidies. Reaching unsubsidized, levelized price parity is the holy grail of the renewable energy world, and recent market trends suggest that reality to be quickly approaching.

However, due to many applications of solar and uneven distribution of solar radiation as a resource, price parity can mean something very different depending on the conversation. This has resulted in a tower of Babel that is likely holding back the growth of the solar industry. In order to navigate the complicated solar economic landscape, it is helpful to outline the key considerations for cost comparisons.

While it may seem like an obvious point, defining the comparison set and identifying the competition is a necessary component to the analysis. There are three main components to this process: (1) ownership, (2) fuel source, and (3) capacity and infrastructure.

The first and most important consideration is whether the system is owned by an entity that pays retail or wholesale price for energy. Retail electric prices can range from two to three times wholesale prices. As a result, while the cost-per-kilowatt hour is higher for smaller, distributed generation systems, the higher price point of retail electricity is dramatically higher. Systems competing with retail electric prices achieve price parity (and beyond) much sooner than those competing with wholesale prices.

There are several reasons for this discrepancy. First and most important, owners of systems that compete with retail electricity are commercial and residential distributed generation systems. These systems have virtually no maintenance or overhead. They have nearly no electron loss because the distance from panel to inverter is

Solar Energy Markets. DOI: http://dx.doi.org/10.1016/B978-0-12-397174-6.00009-X

measured in feet rather than miles. There are no emergency funds required for storm cleanups. And finally, there is no volatility in fuel prices, because it remains a constant zero.

Utility-scale solar systems, while benefitting from efficiencies of scale, are still not cost competitive because they suffer the same inefficiencies as traditional power plants. The utility-scale PV systems therefore need to produce power much more cheaply to compete with fossil fuels.

The difference between wholesale and retail electric price parity is the first opportunity for significant policy crosstalk. At the same time, failing to understand the difference minimizes the incredible potential for affordable, distributed energy production in a variety of markets.

The fuel source(s) is also an important consideration. Fossil fuels tend to have volatile prices that fluctuate based on supply, technological advances, and economic growth that fuels demand. Speculation is also a noted price driver. With such a wide array of global fuel sources and a price structure that varies dramatically, analysis of solar competitiveness should factor in the fixed price of solar production over the life of the system.

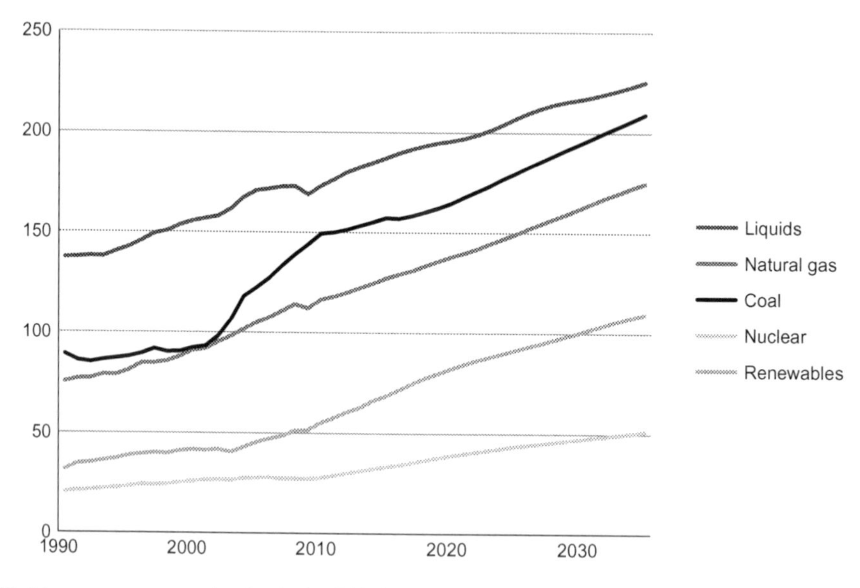

World energy consumption by fuel, 1990-2035 (quadrillion Btu)
Source: U.S. Energy Information Administration International Energy Outlook 2011 DOE/EIA-0484(2011) September 2011 From: http://www.eia.gov/forecasts/ieo/world.cfm

Capacity and infrastructure are also important to consider. As referenced previously, whether a system is distributed generation or utility-scale makes a big difference in cost comparisons. At the same time, the current overall infrastructure is important to consider because solar becomes remarkably more cost competitive vis-à-vis fossil-fuel generation when a new plant needs to be constructed as opposed to merely increasing the capacity of existing plants.

Stated differently, the cost of solar plants (whether rooftop distributed or large, utility-scale plants) is much more favorable when compared to the cost of constructing a new gas or coal plant. These comparisons are markedly different if new fossil plants are not operating near capacity and therefore only require more fuel to produce the competitive power.

Massachusetts (which happens to be the home state of the author, and situ for his grid-connected PV system) is an excellent exemplary state to review these issues. It has aging fossil (and nuclear) infrastructure, an old electrical grid, no coal or gas (and therefore an importer of fuel) resources, and is operating near full capacity at peak hours. It also sits in the Northeast of the United States, which is not the first place that would come to mind for solar cost competitiveness, due to its minimal solar resource when compared to the sunny Southwest. In fact, for utility-scale systems, Massachusetts is near the middle of the pack in terms of achieving price parity (it should be noted that with no traditional energy resources in-state, Massachusetts relies on volatile import prices for fuel).

However, as of 2013, Massachusetts distributed generation for residential applications has the shortest payback of any state. This is primarily for three reasons: (1) high cost of retail electricity at 12.7 cents/kWh, (2) a $2000 state rebate, and (3) renewable energy credits, sold in the private market, that can be worth as much as $6600 over the life of an average rooftop system.

This, coupled with third-party financed solar leases with cheap financing rates, results in overwhelmingly competitive pricing. In fact, the prepaid price for 20 years of average, single family electrical usage in Massachusetts can be as low as 4.3 cents/MWh, nearly one-third the retail rate from the utility.

Several more scientific approaches have been conducted that demonstrate the many considerations for comparing solar electricity as a competitive alternative to other sources. These sources include various ways of addressing competitiveness and also address the history that has driven solar—and particularly PV—price competitiveness across the globe.

A recent paper by Bazilian et al. summarizes the dramatic shift that has occurred since 2008 in regards to the price competitiveness of solar PV power.[1] In 2008, module prices were essentially "flat at $3.50–4.00/W despite manufacturers making continuous improvements to technology and scale to reduce their costs."[2] This was reported to be mainly due to shortages of raw materials and fixed price schemes in Europe, which hampered production and restricted competition.[3]

As raw materials, and particularly, silicon, became more readily available (and at a lower cost), the soaring profit margins led to the rapid expansion of manufacturers across the globe. At the same time, technological advances continued to drive down module costs, allowing "manufacturers to drop their prices by 50%, and still make a positive operating margin."[4] Since then, prices have continued to decline dramatically,

[1] Bazilian et al. Re-considering the Economics of Photovoltaic Power. Bloomberg New Energy Finance, 2012.
[2] *id.* at p. 3.
[3] *id.*
[4] *id.*

to a point where most manufacturers are not profitable. In late 2001, module prices fell below $1.00/W for the first time, a significant benchmark for the industry and prices currently sit between $0.85 and $1.01/W.[5]

Cost per watt is an important component to the price of generated solar electricity, but it is only one component. As a result, many analyses rely instead on Levelized Cost of Energy (LCOE). According to Bazilian et al., "LCOE analysis considers costs distributed over the project lifetime and as such supposedly provides a more accurate economic picture, which system operators prefer over a simple capital cost-per-watt calculation."[6] Globally, the LCOE has declined from $0.32/kWh to as low as $0.11 in 2013 (which may include federal and local incentives in the United States).[7] Of important note, this level is below the average kilowatt hour retail price of $0.115 in the United States.[8]

BNEF identifies the following factors as most important for determining LCOE: (1) capital costs, (2) capacity factor, (3) cost of equity, and (4) cost of debt.[9] Despite these varied factors, a recent analysis by Stefan Reichelstein of Stanford University suggests that northern US locations (and also many locations in Europe and across the globe at higher altitudes) will never be competitive due to the capacity factor.[10] This is not supported by BNEF analysis, suggests that the variables are too great to make such predictions.[11]

Already the Emirates Solar Industry Association has demonstrated that across its climate the LCOE for solar is (unsubsidized) $0.15. "At this level, PV is cheaper on a simple LCOE basis than open-cycle peaking units at gas prices at higher than $5.00/MMBtu," and PV has already replaced some of these plants in the United States, including in San Diego, CA.[12] That this is an outlier rather than the norm, despite clear economic advantages in favor of PV facilities, speaks to the confusion and misinformation provided to policy makers.

It is interesting to note that the term typically used to describe competitiveness, "grid parity," appears to be out of favor because it further confuses the issue. According to Bazilian et al., the term comes from a time when solar was an "underdog" and no longer has much utility for real world decision making.[13] It is specifically problematic because it "does not take into account the value of solar PV to the broader electrical industry, and is often used to compare a retail technology against a wholesale price."[14] Of course, this again addresses the key point about the utility of solar, in that it need not be driven by utilities, but that the retail sector is where it is likely to be most competitive.

[5] *id.* at p. 4.

[6] *id.* at p. 7.

[7] *id.* at p. 8.

[8] http://www.forbes.com/sites/justingerdes/2012/05/24/solar-power-more-competitive-than-decision-makers-or-consumers-realize/.

[9] Bazilian et al. at p. 9.

[10] https://energy.stanford.edu/sites/default/files/StefenReichelstein.pdf.

[11] Bazilian et al. at p. 10.

[12] *id.*

[13] *id.* at p. 13.

[14] *id.*

Much has been made of Chinese production and negative operating margins that have been supported by government subsidies. While the Chinese actions have clearly had some impact on these lower prices, the cost curve resulting from greater efficiency (often called the learning curve) has been even more important. This is critical because it means that while prices are not likely to decline as rapidly as they have in the past, they are expected to continue to decline by 10–20% each year for the next decade.

It is not just traditional Chinese (and non-Chinese) PV panels that will decline in price. The NREL and the US Photovoltaic Manufacturing Consortium (PVMC) have partnered to find ways to reduce the cost and enhance the competitiveness of US thin film based in copper–indium–gallium–sulfide. The partnership is an important part of PVMC goal of reducing the installed price of thin film solar energy systems by 75% by driving down costs in manufacturing.

The partnership, which is funded in part by funds from the SunShot Initiative, includes major research institutions including SUNY-Albany, as well as Sematech, a consortium that covers 50% of the worldwide chip market.[15] Importantly, the research will not only seek to drive down manufacturing costs of current technologies, but to spur newer, cheaper forms of solar.

While manufacturing price declines sound like a boon to the solar industry, it is not the module price declines that will be significant because the module cost only makes up about 20% of current system costs. Other considerations include labor efficiency (in the installation process), streamlined permitting, and other components that will need to be addressed in mature markets to drive down prices.

One of the primary issues in the United States is lowering some of the nonmodule price restrictions in the proliferation of third-party financing and solar leases. According to the SEIA and GTM Research, the first two quarters of 2012 illustrate a trend where third-party leases make up 70–80% of the residential market in states where it is allowed. This success led to the attraction of over $600 million in investments in early 2012 that help to fund the booming installations.[16]

Third-party systems offer flexibility and cost savings to consumers, frequently allowing for low- to no-upfront cost solar installations, or favorable prepayment options for consumers. The low cost of these systems is driven by all of the same components as a purchase system (labor cost and efficiency, module prices, consumer incentives, State Renewable Energy Credits (SRECs), etc.) and also a few additional items. First, because the systems are owned and operated by companies rather than individuals, the depreciation of the system is tax deductible. Allowance for deductions as well as accelerated depreciation has been a key factor to the proliferation of the systems as well as low financing costs driven by historically low interest rates.

The current structure and trends in the industry illustrate a rapidly approaching cost competitiveness in many high-solar resource locations. However, much of the optimism in the sector stems from the potential of developing markets, including several new markets in the Middle East, Africa, and India. Where the traditional source of energy is diesel-electric production, solar PV is already cost competitive.[17]

[15]Cleantechnica, March 15, 2013. Solar R&D heavyweights join in effort to drive down thin-film CIGS costs.
[16]SEIA and GTM Research, 2012. U.S. Solar Market Insight Report: Q2 2012 Executive Summary, p. 3.
[17]Bazilian et al. at pp. 13–14.

Most of the information regarding potential future scenarios can be found in Chapter 9, however, several global trends are important to consider when thinking of the continued competitiveness of solar energy. As most experts and industry professionals expect solar subsidies to decline in the future, the following must occur for solar to keep on its current trend towards cost competitiveness:

1. Module prices must continue to decline. Most experts believe this will occur in the 10–20% per year range, mostly gained from increased efficiency.
2. Developed markets must continue to gain efficiency in their installation sectors. Specifically, the United States should start to look more like Europe in terms of installation efficiency.
3. China should increase its installed capacity at a measured pace to spur the market without causing a run-on panels and spiking prices.
4. Emerging markets should continue to install solar capacity, especially for grid-connected systems to replace diesel production.
5. Governments should continue supporting the industry over the short term (5–10 years) to ensure steady growth until solar reaches scale.

Even with these factors, several other issues are of critical importance to determining the economic competitiveness of solar electricity. These include:

1. *Cost of energy*. This is fairly obvious as it is the point of reference and comparison for the industry. It is particularly important for solar because commodity prices tend to *increase* over time, whereas cost of technology—such as solar panels—tends to *decrease* over time. As a result, the falling fixed cost of solar electric systems—with no required fuel inputs—will become more competitive over time if fossil-fuel prices increase or stay the same. If they decrease, as may be happening in the US natural gas market, price parity may be delayed further.
2. *Pricing peak power*. Many utilities are considering tiered pricing for electricity, to increase the cost of peak power consumption. Because these tend to be daylight hours during the work week (and more typically in the summer months), solar is most productive during peak power periods. Again, comparisons to a more expensive power source are favorable to solar energy, so as a peak power source, solar will become competitive more quickly. In the Southwestern United States, peak power price competitiveness is achieved at $0.15–0.50/kWh and maximum installed cost of $3.25–6.00/W.[18]
3. *Intermittent wholesale power based on fossil-fuel avoidance*.[19] Solar electricity's use as an "opportunistic supply source… that imposes little or no grid integration costs" allows utilities to even out fluctuations on fuel prices.[20] While this attribute is a key component to solar's competitiveness, it diminishes as solar output rises. This is because large-scale solar capacity addition requires the very grid integration that the lower levels do not. Because of solar's peak power performance, it is hard to address this component in a vacuum, but solar is most competitive in this space at $0.035–0.04/kWh and $0.81/W.[21]
4. *Storage*. Ultimately, solar will be limited as a major power source without effective storage. While it can continue to compete with traditional power sources for peak loads, only with storage can it effectively *replace* traditional power systems.

[18]McDonnell, P. Making sense of PV parity, quantifying solar's competitiveness. Renewable Energy World, April 23, 2013, p. 3.
[19]*id*. at p. 1.
[20]*id*. at p. 2.
[21]*id*. at p. 4.

Utility-scale PV is currently not cost competitive with "build-new" gas and coal-fired plants. However, the steep declines in prices and market volatility in the fossil sector suggest that such systems will—in markets with carbon pricing schemes—be competitive by 2020, and on par with high-quality wind power prices.[22] Further declines are nearly certain, all but guaranteeing a much more competitive marketplace in the future.

Solar is rapidly approaching cost competitiveness in a variety of applications. As it gains in competitive advantage, more and more countries will increase solar capacity. Chapter 8 reviews the outlook for several important solar countries.

[22] Bloomberg New Energy Finance. Global Renewable Energy Market Outlook 2013, Fact Pack, April 26, 2013.

10 Afterward

The solar industry has grown in fits and starts since the early 1970s, but its recent ascent has all but cemented it as a major energy technology for the future. In the short and medium term, the future of solar is brightest for installation companies, which are benefitting from low module prices, cheap capital, and abundant new markets. The manufacturing market will remain depressed as the oversupply of panels comes down.

China will play a pivotal role in the future of solar energy and is likely to become the dominant market driver of the industry. It has already eclipsed the United States as the top investment market for solar, but the most important trends are whether it will continue its aggressive shift away from production or towards installation.

China should make efforts towards building a more balanced and sustainable market for solar products, if for no other reason than to protect its investments in the industry. After years of massive subsidies, the Chinese solar production engine has become the market leader but is also significantly oversupplying the market with panels. China, with its immense scale, intense energy requirements, and ability to green-light projects quicker than most countries, could swallow the entire oversupply in a single gulp.

The results of such a movement would be mixed at first. As supply dwindled, prices would rise in the short term, putting pressure on installers in Europe and America. However, the scale of the market would increase to a point where pricing would become both low and sustainable.

Manufacturers would see profits return. This would spur more production innovation and efficiency. It would also mark the return of R&D and further technological innovations to develop the next generation of solar systems.

Policy support will need to remain constant during the transition. In the United States in particular, a continuation of the 30% consumer tax credit, expansion of FiT laws and third-party leases, and continued R&D funding will be critical to maintain the market in flux. In Europe, moving from austerity- to growth-driven policies will be important as that bloc reaches saturation of its solar markets.

Policies will become easier to enact over time, as the shift in public opinion that is underway is both growing in momentum and getting older. In the developed world, and particularly in the United States, political views are still important regarding the perception of clean energy, but much less so in the past. Climate deniers are seen increasingly as outcasts, and the younger generation of Americans tends to balance profits with environmental protection with a longer view in mind.

Without any major setbacks, solar is primed to compete with other energy sources. As their prices rise as supply of fossil fuels dwindle in the face of increasing demand, the price of solar modules continues to fall. And the fuel is forever free. Given current trends, price parity, and then subsidy-free price parity will come sooner than most expect, sometime between 2015 and 2025.

Solar Energy Markets. DOI: http://dx.doi.org/10.1016/B978-0-12-397174-6.00010-6

At these prices, solar systems, particularly PV systems, will dot the landscapes and rooftops of areas with strong solar resources. Jobs will be created by the tens of thousands in the United States, which will be important to offset the continuing declines of the coal industry. Solid, middle-class jobs in manufacturing, construction, sales, and engineering will proliferate in the Middle East, India, and throughout Asia. Massive amounts of CO_2 will be diverted.

As solar power enters the mainstream, the energy future of many nations hangs in the balance. This future increasingly appears clear and bright, and while solar energy will not become a predominant source in the near term, its importance as a supplemental and incremental component will only increase over time.

Glossary

Absorber: The layers used to absorb sunlight.

Acreage: The total acres used up by a photovoltaic system.

Activated shelf life: The time it takes for a charged battery to reach an unusable level when stored at a specific temperature.

Activation voltage: The voltage that the controller will operate at to protect the batteries.

Air mass: In terms of solar energy, the air mass relates to the path length of solar radiation through the atmosphere. For instance, an air mass of 1.0 means that the sun is directly overhead and the radiation travels through one atmosphere of thickness.

Alternating current (AC): Electrical current that constantly reverses direction of flow.

Ambient temperature: The air temperature of the surrounding area.

Ammeter: A device used to measure current flow.

Amorphous silicon: A thin film solar photovoltaic material that has a glassy structure.

Array: Solar modules connected together to form a single structure.

Array current: The electrical current output when a photovoltaic array is exposed to sunlight.

Array operating voltage: The voltage of a photovoltaic array when exposed to sunlight and feeding a load.

Autonomous system: A photovoltaic system that operates independent of any other energy-generating source.

Azimuth: The angle between the north direction and the projection of the array surface into the horizontal plane, measured clockwise from north. A due south facing array would be 180° azimuth.

Balance of system (BOS): All of the components of a photovoltaic system minus the solar module.

Battery capacity: The total number of ampere-hours (Ah) that a charged battery can output.

British Thermal Unit (BTU): The amount of heat energy it takes to raise one pound of water from a temperature of 60°F to 61°F at 1 atmosphere pressure.

Cell: The basic unit of a photovoltaic module or a battery. The cell contains the necessary materials to produce electricity.

Cell efficiency: The ratio of electrical energy produced by a photovoltaic cell to the energy contained in the sunlight that reaches the photovoltaic cell.

Cell junction: The area of contact between two layers (positive and negative) of a photovoltaic cell.

Cloud enhancement: The increase in solar intensity from reflected light due to nearby clouds.

Cogeneration: The joint production of electrical and heat energy at a single location, resulting in a more efficient use of thermal energy.

Concentrator: A photovoltaic device that uses optical elements (mirrors, lenses, etc.) to increase the amount of light that reaches a solar cell. Concentrators track the sun and reflect or enhance only the direct sunlight.

Crystalline silicon: Photovoltaic cell material that is made from a single crystal or polycrystalline ingot of silicon.

Days of storage: The number of days a stand-alone system will power a specified load without any solar energy input.

Dealer: A retailer of photovoltaic products and/or systems.

Design month: The month in which the amount of insolation and load requires the maximum energy from the array.

Developer segment: Types of developers—utility, government, commercial, or residential.

Diffuse insolation: The incident sunlight received from scattering due to obstructions in the atmosphere (clouds, fog, dust, etc.).

Diffuse radiation: The incident radiation received from the sun after reflection and scattering due to obstructions in the atmosphere (clouds, fog, dust, etc.) and on the ground.

Direct insolation: The sunlight falling directly on a collector.

Direct radiation: The light that has traveled in a straight path from the sun (also known as direct beam radiation).

Disconnect: The switch used to connect or disconnect the different components of a photovoltaic system.

Distributed systems: Electrical systems that are installed at or near the locations that the energy will be used. Residential photovoltaic systems are distributed systems.

Distributor: A wholesaler of photovoltaic products.

Downtime: The time when a photovoltaic system cannot provide power to the load. Usually measured as hours per year or as a percentage.

DSSC: Dye-sensitized solar cell.

Duty cycle: The ratio of active to total time for a photovoltaic system.

EG or EGS: Electronic grade silicon.

EVA: An encapsulant used between the glass and solar cells in photovoltaic modules.

Feed-in tariff: An economic policy that is created to promote active investment in and production of renewable energy sources.

Fixed tilt array: A solar photovoltaic array set at an angle to the horizontal.

Flat-plate PV: A photovoltaic array that does not contain concentrating devices and therefore responds to direct and diffuse sunlight.

Fresnel lens: A concentrating lens that is positioned above and is concave to photovoltaic material in order to direct light onto the material.

Gen 3: Third-generation solar cells.

Gigawatt (GW): A measurement of power equal to a 1000 million watts.

Gigawatt-hour (GWh): A measurement of energy. One gigawatt-hour is equal to 1 gigawatt being used over the course of an hour.

Grid-connected: An energy producing system that is connected to the utility transmission grid (grid-tied).

HE: High efficiency.

HF: High reflector.

High-voltage disconnect: The voltage level for which the controller will disconnect the array to prevent the batteries from overcharging.

Hybrid system: A photovoltaic system that also includes some other electricity generating power source.

Incentive programs: Incentive program that the solar project is entitled to.

Incident light: The light that shines on the surface of a photovoltaic cell or module.

Independent power system: An energy generation system that is independent of the main power grid.

Insolation: The amount of sunlight that reaches an area, usually measured in watt hours per square meter per day.

Installers: Market and customize photovoltaic systems for installation.

Integrator: Combines photovoltaic components into a complete system.

Inverter: Converts the DC power from the photovoltaic array to AC power.

Irradiance: The solar power incident on a surface (kilowatts per square meter). Irradiance multiplied by time gives insolation.

JDA: Joint development agreement.

Joule (J): The energy conveyed by 1 watt of power for 1 s.

Junction diode: Solar cells are junction diodes in that they are semiconductor devices that have a junction and a built-in potential that passes current better in one direction than the other.

Kilowatt (kW): A unit of electrical power equal to 1000 watts.

Kilowatt-hour: A unit that describes the amount of energy that derives from a power of 1 kilowatt acting over the period of 1 h.

kWh output: The estimated first-year kilowatt hour output of a photovoltaic system.

Langley: A unit of solar irradiance; 1 calorie per square centimeter.

Light trapping: The trapping of light inside a semiconductor material by reflecting and refracting the light at critical angles.

Load: The electrical power that is consumed at any given moment or averaged over a specified period. The load that an electric generating system supplies varies greatly by time of day and to an extent, by season or time of year.

Low-voltage disconnect (LVD): The voltage level at which the controller will disconnect the load from the batteries to prevent over-discharging.

Market: The quantity of modules delivered to final photovoltaic installation sites, which includes modules awaiting installation or grid connection.

mc-Si: Multicrystalline silicon.

Megawatt (MW): A measurement of power that is equal to 1 million watts.

Megawatt-hour (MWh): A measurement of power that incorporates time. One megawatt-hour is equal to 1 megawatt being used for a period of 1 h.

MG-Si: Metallurgical grade silicon.

Microgroove: A small groove on the surface of a cell that can be filled with metal for contacts.

Module: An encapsulated panel that contains electrically connected photovoltaic cells.

Monocrystalline solar cell (mono-Si): A type of solar cell made from a thin slice of a single large crystal of silicon.

Multijunction device: A photovoltaic device containing one or more cell junctions that may be different in nature but are optimized to absorb particular parts of the solar spectrum in order to achieve higher overall cell efficiency.

Net metering: The practice of exporting surplus solar power to the electricity grid. The homeowner electric meter either physically moves backwards or a financial credit will be applied to the homeowner's electric bill.

Normal operating cell temperature (NOCT): The estimated temperature of a solar photovoltaic module when it is in operation.

One-axis tracking: A photovoltaic system structure that rotates on a single axis in order to track the movement of the sun.

Panel: Used interchangeably with "module."

Passive solar: Utilizing part of a building as solar collector, as opposed to active solar, such as photovoltaics.

Peak sun hours: The number of hours per day when solar irradiance averages $1000\,W/m^2$.
Photovoltaic (PV): Any device that produces free electrons when exposed to light.
Photovoltaic array: Photovoltaic modules connected together in a single structure.
Photovoltaic cell: The smallest discrete part of a photovoltaic module that is responsible for the conversion of light into electrical energy.
Polycrystalline silicon: A material used to make photovoltaic cells which consists of many crystals.
Power purchase agreement (PPA): When a solar company covers the full cost of installing and maintaining a solar system. In return, the host customer agrees to buy the power produced by the system.
Pyranometer: An instrument used for measuring global solar irradiance.
Semiconductor: A material that has an electrical conductivity in between that of a metal and an insulator. Semiconductors for photovoltaic cells include silicon, gallium arsenide, copper indium diselenide, and cadmium telluride.
Solar energy: Electromagnetic energy transmitted from the sun. The energy must be captured and converted to AC electrical power.
Solar thermal: A form of power generation that uses concentrated sunlight to heat fluid that may be used to drive a motor or turbine.
Substrate: The material on which a photovoltaic cell is made.
Superstrate: The covering on the sun side of a photovoltaic module providing protection from environmental hazards.
System operating voltage: The output voltage of a photovoltaic array under load.
Thin film: A thin layer of semiconductor material, generally a few microns or less, used to make photovoltaic cells. Examples include copper indium diselenide, cadmium telluride, gallium arsenide, and amorphous silicon.
Tilt angle: The angle at which a solar array is tilted towards the sun.
Two-axis tracking: A photovoltaic array tracking system that is able to rotate independently about two axes (vertical and horizontal).
UMG: Upgraded metallurgical grade silicon.
Volt (V): The amount of force required to drive a steady current.
Wafer: A thin sheet of photovoltaic material made from cutting from a single crystal or ingot.
Watt (W): The standard unit of measure for power either for capacity or demand.
Zenith angle: The angle between the sun (the point of interest) and the zenith (directly overhead).